Randomization in
Clinical Trials

Randomization in Clinical Trials

Theory and Practice

WILLIAM F. ROSENBERGER
University of Maryland, Baltimore County

JOHN M. LACHIN
The George Washington University

A JOHN WILEY & SONS, INC., PUBLICATION

This text is printed on acid-free paper. ∞

Copyright © 2002 by John Wiley & Sons, Inc., New York. All rights reserved.

Published simultaneously in Canada.

No part of this publication may be reproduced, stored in a retrieval system or transmitted in any form or by any means, electronic, mechanical, photocopying, recording, scanning or otherwise, except as permitted under Section 107 or 108 of the 1976 United States Copyright Act, without either the prior written permission of the Publisher, or authorization through payment of the appropriate per-copy fee to the Copyright Clearance Center, 222 Rosewood Drive, Danvers, MA 01923, (978) 750-8400, fax (978) 750-4744. Requests to the Publisher for permission should be addressed to the Permissions Department, John Wiley & Sons, Inc., 605 Third Avenue, New York, NY 10158-0012, (212) 850-6011, fax (212) 850-6008, E-Mail: PERMREQ @ WILEY.COM.

For ordering and customer service, call 1-800-CALL WILEY.

Library of Congress Cataloging-in-Publication Data is available.

ISBN 0-471-23626-8

Printed in the United States of America

10 9 8 7 6 5 4 3 2 1

Contents

Preface

The Department of Statistics at The George Washington University (GWU) was a hotbed of activity in randomization during the 1980s. L. J. Wei was on the faculty during the early eighties, and drew Bob Smythe into his randomization research with some interesting asymptotics problems. At the same time, John Lachin was working on his series of papers on randomization for *Controlled Clinical Trials* that appeared in 1988. He, too, was influenced by Wei, and began advocating the use of Wei's urn design for clinical trials at The Biostatistics Center, which he directed at that time, and now co-directs. I studied at GWU from 1986–1992, taking many classes from Lachin, Smythe, and also the late biostatistician Sam Greenhouse. I wrote my doctoral thesis under the direction of Smythe, on asymptotic properties of permutation tests and response-adaptive randomization, topics covered in the latter chapters of this book. I also worked on clinical trials at The Biostatistics Center from 1990–1995 under the great clinical trialist Ray Bain (now at Merck). Needless to say, I was well indoctrinated in the importance of randomization to protect against biases, and the importance of incorporating the particular randomization design into analyses.

I currently continue my research on randomization and adaptive designs at University of Maryland, Baltimore County, where I teach several graduate-level courses in biostatistics and serve as a biostatistician for clinical trials data and safety monitoring boards for the the National Institutes of Health, the Veteran's Administration and industry. One of my graduate courses is the design of clinical trials, and much of this book is based on the notes from teaching that course.

The book fills a niche in graduate-level training in biostatistics, because it combines both the applied aspects of randomization in clinical trials along with a probabilistic treatment of properties of randomization. Although the former has been covered in many books (albeit sparsely at times), the latter has not. The book takes an unabashedly non-Bayesian and nonparametric approach to inference, focusing mainly on the linear rank test under a randomization model, with some added discussion on likelihood-based inference as it relates to sufficiency and ancillarity. The strong focus on randomization as a basis for inference is another unique aspect of the book.

Chapters 1–12 represent the primary focus of the book, while Chapters 13–15 present theoretical developments that will be interesting for Ph.D. students in statistics and those conducting theoretical research in randomization. The prerequisites for Chapters 1–12 is a course in probability and mathematical statistics at the advanced undergraduate level. The probability in those chapters is presented at the level of Sheldon Ross's *Introduction to Probability Models*, and a thorough knowledge of only the first three chapters of that book will allow the student to get through the text and problem sets of those chapters (with the exception of Section 3.6, which requires material on Markov chains from Chapter 4 of Ross). Chapters 13–15 require probability at the level of K. L. Chung's *A Course in Probability Theory*. Chapter 13 excerpts the main results needed in large-sample theory for Chapters 14 and 15.

Problem sets are given at the end of each chapter; some are short theoretical exercises, some are short computer simulations that can be done efficiently in SAS, and some are questions that require a lot of thinking on the part of students about ethics and statistical philosophy, and are useful for inspiring discussion. I have found that students love to read some of the great discussion papers on such topics as randomization-based inference, the ECMO controversy, and ethical dilemmas in clinical trials. I try to have two or three debates during a semester's course, in which every student is asked to present and defend a viewpoint. Some students are amazed, for instance, that there is any question about appropriate techniques for inference, because they have been presented a single viewpoint in their mathematical statistical course, and have basically taken their instructor's lecture notes as established fact.

One wonderful side-benefit of teaching randomization is the opportunity to meld the concepts of conditional probability and stochastic processes into real-life applications. Too often probability is taught completely independently of applications, and applications are taught completely independently of probability and statistical theory. As each randomization sequence forms a stochastic process, exploring the properties of randomization is an exercise in exploring the properties of certain stochastic processes. I have used these randomization sequences as illustrations when teaching stochastic processes.

This book can be used as a text for a one-quarter or one-semester course in the design of clinical trials. In our one-semester course, I supplement this material with a unit on sequential monitoring of data. I assume that students already have a basic knowledge of survival analysis, including the logrank family of tests and hazard functions. Computational problems can be done in SAS, or in any other programming language, such as MATLAB, but I anticipate students would be facile in SAS before taking such a course.

I also hope that this book will be quite useful for statisticians and clinical trialists working in the pharmaceutical industry. Based on my many conversations and collaborations with statisticians in industry and government, I believe the fairly new techniques of response-adaptive randomization are attractive to industry and also to the Food and Drug Administration. This book will be the first clinical trials book to devote a substantial portion to these techniques. However, this book should not be construed as a book on "adaptive designs". Adaptive design has become a major subdiscipline of experimental design over the past two decades, and the breadth of this subdiscipline make a book on the subject very difficult to write. In this book, we focus on adaptive designs only as they relate to the very narrow area of *randomized* clinical trials.

Finally, the reader will note many "holes" in the book, representing open problems. Many of these concern randomization-based inference for covariate-adaptive and response-adaptive randomization procedures, and also some for more standard restricted randomization, in areas of group sequential monitoring and large sample theory. I hope this book will be a catalyst for further research in these areas.

Acknowledgments: I am grateful for the help and comments of Boris Alemi, Steve Coad, Susan Groshen, Janis Hardwick, Karim Hirji, Kathleen Hoffman, Feifang Hu, Vince Melfi, Connie Page, Anindya Roy, Andrew Rukhin, Bob Smythe, and Thomas Wanner. Yaming Hang researched sections of Chapter 14 during a one-year research assistantship. During the writing of this book, I was supported by generous grants from the National Institute of Diabetes and Digestive and Kidney Diseases and the National Cancer Institute. Large portions of the book were written during the first semester of my sabbatical spent at The EMMES Corporation, a clinical trials coordinating center in Rockville, MD. I am grateful to EMMES, in particular Ravinder Anand, Anne Lindblad, and Don Stablein, for their support of this research and their kindness in allowing me to use their office resources. On the second semester of my sabbatical, I was able to "test" a draft of the book while teaching Biostatistics 219 in the Department of Biostatistics, UCLA School of Public Health. I thank Bill Cumberland and Weng Kee Wong for arranging a visiting position there and the students of that course for finding a good number of errors.

W. F. R.

Baltimore, Maryland

I joined the Biostatistics Center of the George Washington University in 1973, one year after receiving my doctorate, to serve as the junior staff statistician for the National Institutes of Health (NIH) funded multi-center National Cooperative Gallstone Study (NCGS). Jerry Cornfield and Larry Shaw were the Director and Co-Director of the Biostatistics Center, and the Principal Investigator and Co-Principal Investigator of the NCGS coordinating center. Among my initial responsibilities for the NCGS were to determine the sample size and to generate the randomization sequences. Since I had not been introduced to these concepts in graduate school, I started with a review of the literature that led to a continuing interest in both topics.

While Jerry Cornfield thought of many problems from a Bayesian perspective, in which randomization is ancillary, he thought that randomization was one of the central characteristics of a clinical trial. In fact he once remarked that the failure of Bayesian theory to provide a statistical justification for randomization was a glaring defect. Thus in 1973–1974, Larry Shaw and I approached the development of randomization for the NCGS with great care. Larry and I agreed that we should employ a procedure as close to complete randomization (toss of a coin) as possible and decided to use a procedure that Larry had previously employed in trials he organized while a member of the Veterans Administration Cooperative Studies Program. That technique has since come to be known as the "big stick" procedure.

Later, around 1980, I served as the Principal Investigator for the statistical co-ordinating centers for the NIH-funded Lupus Nephritis Collaborative Study and the Diabetes Control and Complications Trial. Both were unmasked studies. In the late seventies I first met L. J. Wei while he was on sabbatical leave at the National Cancer Institute. He later joined the faculty at George Washington University and we became close friends and colleagues. Thus when it came time to plan the randomization for these two studies, I was drawn to Wei's urn design because of its many favorable properties. Later, I organized a workshop "The Role of Randomization in Clinical Trials" for the 1986 meeting of the Society for Clinical Trials. The papers from that workshop, co-authored with John Matts and Wei were then published in *Controlled Clinical Trials* in 1988. In 1990–1991 I had a sabbatical leave, during which I began to organize material from these papers and other research into a book.

In 1991–1992 I taught a course on clinical trials in which I used the material from the draft chapters and my 1988 papers. One of the students auditing that course was Bill Rosenberger. Bill was concurrently writing his dissertation on large sample inference for a family of response-adaptive randomization procedures under the direction of Bob Smythe. Bob had conducted research with Wei and others on randomization-based inference for the family of urn designs. Bill went on to establish a strong record of research into the properties of response-adaptive randomization procedures.

In 1998 I again took sabbatical leave that I devoted to the writing of my 2000 text *Biostatistical Methods: The Assessment of Relative Risks*. During that time Bill suggested that we collaborate to write a text on randomization. This book is the result.

In writing this text we have tried to present the statistical theoretical foundation and properties of randomization procedures, and also provide guidance for statistical practice in clinical trials. While the book deals largely with the theory of randomization, we summarize the practical significance of these results throughout, and some chapters are devoted to practical issues alone. Thus we hope this text will be of use to those interested in the statistical theory of the topic, as well as its implementation.

Acknowledgments: I especially wish to thank L. J. Wei and Bob Smythe for their friendship and collaboration over the years, and Naji Younes for assistance. I also wish to thank those many statisticians who worked with me to implement randomization procedures for clinical trials, and the many physicians who collaborated in the conduct of these studies. Thank you for vesting the responsibility for these studies with me, and for taking randomization as seriously as do I.

J. M. L.

Rockville, Maryland

1

Randomization and the Clinical Trial

1.1 INTRODUCTION

The goal of any scientific activity is the acquisition of new knowledge. In empirical scientific research, new knowledge or scientific results are generated by an investigation or study. The validity of any scientific results depends on the manner in which the data or observations are collected, *i.e.*, on the design and conduct of the study, as well as the manner in which the data are analyzed. Such considerations are often the areas of expertise of the statistician. Statistical analysis alone is not sufficient to provide scientific validity, because the quality of any information derived from a data analysis is principally determined by the quality of the data itself. Therefore, in the effort to acquire scientifically valid information, one must consider all aspects of a study: design, execution, and analysis.

This book is devoted to a time-tested design for the acquisition of scientifically valid information – the randomization of study units to receive one of the study treatments. One can trace the roots of the randomization principle to Sir R. A. Fisher (*e.g.*, 1935), the founder of modern statistics, in the context of assigning "treatments" to blocks or plots of land in agricultural experiments. The principle of randomization is now a fundamental feature of the scientific method and is employed in many fields of empirical research. Much of the theoretical research into the principles and properties of randomization has been conducted in the domain of its application to *clinical trials*. A clinical trial is basically an experiment designed to evaluate the beneficial and adverse effects of a new medical treatment or intervention. In a clinical trial, often subjects sequentially enter a study and are randomized to one of two or more study treatments. Clinical trials in medicine differ in many respects from randomized experiments in other disciplines, and clinical trials in humans involve complex ethical issues which are not encountered in other scientific experiments. The use of randomization in clinical trials has not been without controversy, as we shall see, and statistical issues for randomized clinical trials can be very different from those in other types of studies. Thus this book shall address randomization in the context of clinical trials.

Randomization is an issue in each of the three components of a clinical trial: design, conduct, and analysis. This book will deal with all three elements; however, we will focus principally on the statistical aspects of randomization in the clinical trial, which are applied in the design and analysis phases. Other, more general books are available on the proper conduct of clinical trials [see, for example, Tygstrup, Lachin, and Juhl (1982), Buyse, Staquet, and Sylvester (1984), Pocock (1984), Piantadosi (1997), Friedman, Furberg, and DeMets (1998), Chow and Liu (1998), Matthews (2000)]. These references also give a less detailed development of randomization.

1.2 CAUSATION AND ASSOCIATION

Empirical science consists of a body of three broad classes of knowledge: descriptions of phenomena in terms of observable characteristics of elements or events; descriptions of associations among phenomena; and, at the highest level, descriptions of causal relationships between phenomena. The various sciences can be distinguished by the degree to which each contains knowledge of the three classes. For example, physics and chemistry contain large bodies of knowledge on causal relationships. *Epidemiology*, the study of disease incidence, its risk factors, and its prevention, contains large bodies of knowledge on phenomenologic and associative relationships. Although a major goal of epidemiologists is to determine causative relationships, for example, causal relationships between risk factors and disease that can potentially lead to disease prevention, the leap from association to causation is a difficult one. Jerome Cornfield's (1959) treatise on "Principles of Research" gives a beautifully written account of the history of biomedical studies and the emergence of principles underlying epidemiological research.

Cornfield points to a mass inoculation against tuberculosis in Lubeck, Germany, in 1926. A ghastly episode occurred where 249 babies were accidentally inoculated with large numbers of virulent bacilli. In a follow-up of those babies, 76 had died, but 173 were still free of tuberculosis when observed 12 years later. If the tuberculosis bacilli cause tuberculosis, why didn't all the children develop the disease? The answer, of course, is the dramatic variability in human response to even large doses of a deadly agent. Thus, as we all know, tuberculosis bacilli cause tuberculosis, but *causation* in such cases does not mean that all those exposed to a pathogen will experience the ill effects.

Similarly, one can ask the famous question, why doesn't everyone who smokes develop lung cancer? One possible answer that would please the tobacco industry is that there is a hormonal imbalance that both causes lung cancer and causes an insatiable craving for cigarettes. An alternative answer is that there are *competing risks*: something else kills them first. The most probable answer is that not all those who smoke will develop cancer, due to biological or genetic variation.

Humans have such a complex and varied physiology; they are exposed to so many different environmental conditions; their health is also deeply tied to complex mental states. How can a scientist possibly sift through all the associations one can find between health and these other factors to find causes or cures for disease? One of the oldest principles of scientific investigation is that new information is obtained from a comparison of alternate states. Thus a *controlled clinical trial* is an experiment designed to determine if a medical innovation (*e.g.*, therapy, procedure, or intervention) alters the course of a disease by comparing the results of those undertaking the innovation with those of a group of subjects not undertaking the innovation.

Perhaps the first comparative study of record is the biblical account of Daniel (Chapter 1) in approximately 605 B.C.E., on the effects of a vegetarian diet on the health of Israelites. Rather than be placed on the royal diet of food and wine of the Babylonian court, Daniel requested that his people be placed on a diet of vegetables.

> 'Test us for ten days,' he said, '...then compare us with the young men who are eating the food of the royal court, and base your decision on how we look....'
> When the time was up, they looked healthier and stronger than all those who had been eating the royal food.

Another famous example of a controlled intervention study is Lind's account of the effects of different elixirs on scurvy among British seamen in 1753. His study showed the beneficial effects of citrus, and led (50 years after the study) to the Royal Navy's decision to store citrus on long voyages.

While the idea of comparing those on the innovative treatment with a control group sounds obvious to us today, historically it was not always entirely clear whom to include in the innovation and control groups. At the turn of the twentieth century, an anti-typhoid inoculation movement created controversy between Sir Almroth Wright, a famous immunologist, and Karl Pearson, who, along with Fisher, was a founder of modern statistics. Sir Wright gave the inoculation to anyone who wanted it and compared the subsequent incidence of typhoid with a group of men who refused the inoculation. Here is Pearson's first writing on the subject (Cornfield (1959, pp. 244–245)):

> Assuming that the inoculation is not more than a temporary inconvenience it would seem possible to call for volunteers, but while keeping a register of all men who volunteered only to inoculate every second volunteer. In this way any spurious effect really resulting from a correlation between immunity and caution would be got rid of.

Four years later, Pearson's opinion was even stronger:

> Further the so-called controls cannot be considered true controls, until it is demonstrated that the men who are most anxious and particular about their own

> health, the men who are most likely to be cautious and run no risk, are not the very men who will volunteer to be inoculated.... Clearly what is needed is the inoculation of one half only of the volunteers, equal age incidence being maintained if we are to have a real control.

Pearson recognized what the immunologist did not: that human response to infectious, preventive, or therapeutic agents is variable and is positively related to patient characteristics, such as a willingness to volunteer to receive a new treatment. Thus positive steps must be taken in the design and conduct of a study to eliminate sources of incomparability between those treated and the controls. The inoculated group cannot be compared to any arbitrary control group. The control group must be comparable to the treated group with respect to immune background, hygiene, age, etc. Such factors are called *confounding variables*, because incomparability of the groups with respect to any such factors may confound the results and influence the answer to the research hypothesis.

These considerations play a major role in the design, conduct, and analysis of epidemiologic studies today. In an *observational epidemiologic study*, naturally occurring populations are studied to identify factors associated with some outcome. Since such studies do not employ a randomized design, the results are subject to various types of bias [*cf.* Breslow and Day (1980, 1987), Rosenbaum (1995), Selvin (1996), Kelsey, Whittemore, Evans, *et al.* (1996), among others]. In a retrospective study, these populations consist of *cases* that develop the disease and controls that do not, so that a direct comparison can be made. Just as Pearson noted that there should be equal age incidence in both the inoculated and control groups, epidemiologists may also use *matching* on important variables (*covariates* or *prognostic factors*) that may confound the outcome. Matching is usually done, for instance, on important demographic factors, such as gender, age, and race. Each "case subject" will have a "control subject" with similar characteristics on matched covariates. This allows for greater comparability between the comparison groups. However, it is impossible to match on all known covariates that may influence outcome. Therefore, the leap from association to causation is again tenuous.

The most famous epidemiologic studies were those that demonstrated that smoking causes lung cancer. In 1964, the Report of the Advisory Committee to the Surgeon General was issued that led to warning labels on cigarette packages and restrictions on advertising. The report summarized the evidence from numerous studies that had shown an association between smoking and increased risk of lung cancer and other diseases. Despite any randomized controlled experiments, and based only on observational studies, the Committee concluded that the epidemiologic evidence showed that smoking was indeed a cause of lung cancer. The establishment of a causal relationship between tobacco smoking and cancer created much controversy (and does to this day in some circles). The Surgeon General's report on "The Health Consequences of Smoking" clarified the issue with a definitive statement on what types of evidence from observational studies can lead to a determination of a causal relationship. The Committee (1982, p. 17) stated:

The causal significance of an association is a matter of judgment which goes beyond any statement of statistical probability (*sic*).... An entire body of data must exist to satisfy specific criteria;... when a scientific judgment is made that all plausible confounding variables have been considered, an association may be considered to be direct (causal)....

The Committee stated that the following five criteria must be satisfied:

1. *Consistency of the association.* Diverse methods of approach should provide similar conclusions. The association should be found in replicated experiments performed by different investigators, in different locations and situations, at different times, and using different study methods.

2. *Strength of the association.* Measures of association (*e.g.*, relative risk, mortality ratio) should be large, indicating a strong relationship between the etiologic agent and the disease.

3. *Specificity of the association.* Specificity refers to the precision with which one component of an associated pair predicts the occurrence of the other component in the same individual. For instance, how precisely will smoking predict the occurrence of cancer in an individual? The researcher must consider that agents may be associated with multiple diseases and that diseases may have multiple causes. A single naturally occurring substance in the environment may cause the disease. A single factor can also be a vehicle for several different substances (*e.g.*, tar and nicotine in tobacco), and these may have synergistic or antagonistic effects. There is also no reason to believe that one factor has the same relationship with a different disease with which it is associated. For example, smoking is also associated with heart disease, but perhaps in conjunction with dietary factors that are not important in lung cancer.

4. *Temporal relationship of the association.* Exposure to the etiologic agent must always precede the disease.

5. *Coherence of the association.* The association must make sense in light of our knowledge of the biology and natural history of the disease.

The nine largest studies cited in the Surgeon General's report comprised almost 2 million patients with 17.5 million patient-years of exposure. Based on these data and the convergence of evidence from other sources, one can be confident that smoking "causes" lung cancer, even though the precise causal agent has not been identified (*i.e.*, tar, nicotine, or other agents), and even though no randomized experiment of the effects of smoking and lung cancer has ever been performed.

The overriding question in determining causality in such instances is whether the design or analysis has controlled or "adjusted" for all possible extraneous variables that might account for higher incidence of the disease. Some would say that only a randomized study can ensure adequate control for such factors. It is instructive to note that Fisher, the father of randomization, was never convinced of the link between smoking and lung cancer, and perhaps equally instructive to note that he was a dedicated smoker. Today, almost all epidemiologists and biostatisticians will accept consistent, replicated, careful observational evidence, and few would argue the potency of the evidence against tobacco.

However, it is rare that an adequate body of evidence is amassed from epidemiologic studies alone to assert the above conditions. The number of studies, patients, and extent of exposure required to establish a definite cause by epidemiologic investigation is far greater, and the results ultimately less compelling, than those obtained from a randomized clinical trial, when such trials are possible.

1.3   RANDOMIZED CLINICAL TRIALS

In this book, we will refer to clinical trials which are prospective comparisons of two or more treatments, one or more of which is a new innovation under test, and one or more of which is a control. The most common is a *therapeutic trial*, in which a new therapy, such as a pharmaceutical agent (drug) is compared to a conventional therapy. In a *placebo-controlled* clinical trial of a new pharmaceutical agent, a group of drug-treated subjects may be compared to a group who receive a placebo control [a placebo being a drug preparation (*e.g.*, pill) that is identical to the active therapy, but with inert (inactive) ingredients]. When an established therapy already exists, the new drug may be compared to an *active control*, where the control group receives the established therapy. Therapeutic pharmaceutical clinical trials are often called *phase III* clinical trials, because they represent the third phase of a four-phase process in investigating a promising new therapy. From development of a new pharmaceutical agent to its approval, there is often a *phase I* clinical trial, a small trial to determine the potential toxicity of different dose levels of a drug, and a *phase II* clinical trial, a preliminary study of toxicity and efficacy. A *phase IV* clinical trial involves post-approval follow-up of patient status. These phases are particularly seen in the study of cancer chemotherapeutic agents [see Buyse, Staquet, and Sylvester (1984)], and the four-phase process is often streamlined in other specializations of medicine.

The innovation, however, need not be a simple drug. In some cases a new procedure is evaluated. An example is the Lupus Nephritis Collaborative Study that desired to assess the effects of plasmapheresis on the progression of lupus nephritis, or kidney disease associated with lupus erythimatosis (Lewis, Hunsicker, Lan, *et al.*, 1992). Patients in the plasmapheresis group were hospitalized for a month to undergo daily plasma filtration and exchange, followed by the initiation of standard immunosuppressive therapy consisting of cytoxan and prednisone, the dose of the latter tapered when the patient responded favorably. The patients in the control group received comparable immunosuppresive therapy without initial plasmapheresis.

In other cases a new intervention or an entire treatment regimen of multiple therapies is compared to a control regimen. An example is the Diabetes Control and Complications Trial, which was designed to assess whether a program of intensive therapy aimed at maintaining near-normal levels of blood glucose in subjects with type I diabetes mellitus would prevent or retard the progression of microvascular complications associated with diabetes. Patients in the intensive treatment group received aggressive insulin therapy with frequent monitoring of glucose levels, in conjunction with dietary counseling and exercise. Patients in the conventional treatment group received conventional therapy aimed at maintaining general well-being.

While intensive therapy greatly reduced the risks of complications compared to conventional therapy, such an overall comparison alone cannot identify the mechanism by which the treatment had its effects (Diabetes Control and Complications Trial Research Group, 1993). Subsequent analyses, however, indicated that the effects of intensive treatment were indeed wholly accounted for by the reductions in blood glucose levels.

Some call such trials *pragmatic trials* because the innovation consists of two or more possible agents or procedures used in combination, such that the overall group comparisons alone can not identify the mechanism by which the innovation produces its effects. However, the pragmatist would argue that conclusive evidence that the innovation is indeed beneficial in practice is adequate for its adoption even when the mechanism of the effect is unknown.

The pivotal component of phase III clinical trials is *randomization*, or random assignment of patients to receive either the experimental treatment(s) or control. Cornfield (1959, p. 245) summarized the importance of randomization:

> 1. It controls the probability that the treated and control groups differ more than a calculable amount in their exposure to disease, in immune history, or with respect to any other variable, known or unknown to the experimenter, that may have a bearing on the outcome of the trial. This calculable difference tends to zero as the size of the two groups increase.
> 2. It makes possible, at the end of the trial, the answer to the question "In how many experiments could a difference of this magnitude have arisen by chance alone if the treatment truly has no effect?" It may seem mysterious that a mathematician could actually predict the course of future experiments. All you have to do is compute what would happen if a given set of numbers were randomly allocated in all possible ways between the two groups. Randomization allows this.

The first property of randomization is that it promotes comparability among the study groups. Such comparability can only be attempted in observational studies by adjusting for or matching on *known* covariates, with no guarantee or assurance, even asymptotically, of control for other covariates. Randomization, however, extends a high probability of comparability with respect to unknown important covariates as well. The second property is that the act of randomization provides a probabilistic basis for an inference from the observed results when considered in reference to all possible results. This randomization approach to inference is very different from the usual testing of unknown parameters arising from an independent and identically distributed sample from a known distribution. Later we will deal in detail with these and other precise statistical properties of randomization.

In Cornfield's first point, we come to the root importance of the randomized clinical trial. As scientists are interested in descriptions of phenomena, association among phenomena, and then mechanisms of causation, then the biomedical studies for each require increasing standards of evidence. Basic science research often involves the description of phenomena, observational studies lead to the determination of associations among phenomena, and randomized clinical trials lead to definitive statements on causative effects of agents or regimens on disease processes. As we have seen, despite the fact that consistent, replicated observational studies can also lead us to determine causality, there may always be questions as to whether we have controlled for all factors relating to incidence and prognosis of a disease. The randomized clinical trial allows this control, and hence represents the highest standard of evidence among biomedical studies.

Among the first clinical trials, as we know them today, were the trials performed under the direction of Sir Bradford Hill in the 1940s by the Medical Research Council. These were the first medical trials to employ randomization to individual patients, and constituted a major advance. They led to important findings in many of the persistent diseases of the day, such as whooping cough and tuberculosis. In every respect they were similar to the most rigorous trials conducted today.

The polio vaccine trial of 1954 changed the face of public health worldwide [see Francis (1955)]. Approximately 400,000 children were randomized to receive either the vaccine or a saltwater injection. The results showed a relative risk of 2.5, in favor of the vaccine group. The success of this study belies the controversy among study participants about the need for a controlled, randomized study. In fact, in a quotation attributed to Jonas Salk, it appears Salk was not convinced of the need for a placebo control in the polio trial (source unknown):

> In talks with many people in our own group ... and others as well, I found but one person who rigidly adhered to the idea of a placebo control and he is a bio-statistician who, if he did not adhere to this view, would have had to admit his own purposelessness in life.

In the end, randomized controls were felt necessary because of the variability of incidence of polio from year to year. It was largely due to trials like the polio vaccine trial that convinced the medical community of the value of the randomized clinical trial. Today, it is often considered the "gold standard" among techniques of ascertaining medical evidence.

A good example of the benefits of randomization can be seen in the National Cancer Institute's clinical trial of 62,000 women covered by the Health Insurance Plan (HIP) of Greater New York, commonly known as the HIP trial (see Cairns, 1985). The women were randomized into a "test" group, who were offered a free annual physical examination and mammography for early detection of breast cancer and a "control" group who were given no special encouragement to be examined. The trial was designed to determine if the act of offering free mammography examinations reduces deaths from breast cancer. The results were encouraging. Among the test group, there were 2.9 deaths per 1,000 women in the first nine years, and among the control group there were 4.1 deaths per 1,000 women. The two groups were comparable in their incidence of breast cancer and in terms of general mortality from causes other than cancer, as should be the case because the experiment was randomized. But the results of the trial were also interesting because, among the test group, those who refused examination had a lower death rate due to breast cancer (2.8 per 1,000) than those who accepted the mammography (3.0 per 1,000). This demonstrates the danger of accepting observational data at face value, as one might have concluded that mammography was not effective. The acceptance and rejection groups within the test group were self-selected, and hence subject to confounding due to incomparability with respect to important covariates. In this case, Cairns (1985) believes the confounding variable to be education level. Since better-educated women are known to be more likely to have breast cancer, and less well-educated women are more likely to have less interest in their health, and consequently are more likely to reject examination, the observational component of this study was biased in favor of the rejection group. [For an instructive set of homework problems on the HIP data, see Freedman, Pisani, and Purves (1998), Problems 9 and 10, pp. 22, 23.]

1.4 ETHICS OF RANDOMIZATION

Randomized clinical trials use probability as a method of assigning treatments to patients. Many have argued that probability has no role in medicine, and that only a physician can decide which treatment a patient should receive, using his or her best judgment. However, clinical trials present a unique situation in which new innovations, such as investigational drugs, are being tested for efficacy and safety. Until a drug is proven to be effective and adequately safe, or ineffective or harmful, or just ineffective, the physician is in a state of *equipoise*: a state of genuine uncertainty about which experimental therapy is more effective. Most ethicists would agree, in principle, with the concept that it is ethical to employ randomization in a state of true equipoise, provided the patient consents to be a study participant and is fully informed about the potential benefits and risks of the treatments to be compared in the study.

However, ethics involving human experimentation are seldom so simplistic. On the one hand, a clinical trial gives the patient a chance of being assigned to a potentially beneficial therapy that would not be obtainable elsewhere. But that therapy may also be highly toxic. There is also a chance that a patient will be assigned to a placebo, in effect being denied a therapy that may later prove to be very beneficial (or, on the other hand, harmful). Decisions to enroll in a clinical trial are difficult ones, for this reason, and the patient must often be willing to make a sacrifice for the benefit of our public health.

These considerations exemplify the delicate balance between *individual ethics* and *collective ethics* [see Palmer and Rosenberger (1999)]. Individual ethics dictate what is best for the individual patient, while in collective ethics, we consider the advancement of public health through careful scientific experimentation. In a broad sense, collective ethics leads to individual ethics, as it is only when careful scientific experimentation has yielded a universal standard of care for a given disorder that physicians will be fully informed and will have a scientific basis for the assignment of the best therapy to an individual patient. Although experimentation may lead to many patients being assigned an inferior therapy prior to the determination of the standard of care, this is the price an informed society must pay to obtain the evidence necessary to support informed therapeutic decisions. Such ethical dilemmas are naturally controversial, and are the subject of many treatises and texts (*e.g.*, Engelhardt, 1996).

Some would argue that equipoise is rarely present at the beginning of a phase III clinical trial. Animal studies and phase I and II clinical trials data, plus information on the biological action of the innovation (*e.g.*, drug), combine to create in the mind of many physicians a belief in the effectiveness of one therapy over another. But such confidence may often be premature. The literature is replete with results of negative studies, where promising therapies were shown to be ineffective or even terribly harmful. If equipoise is defined in the confines of a single physician's "hunches" or intuition about a therapy rather than in a global standard of evidence based on randomized controlled studies, there will be no advancement of medical science. This is not to say that careful, replicated, consistent observational studies, as defined in Section 1.2, are not useful and cannot be convincing. But randomization adds an additional component that mitigates contention, and the National Institutes of Health and U. S. Food and Drug Administration now consider a well-conducted, randomized clinical trial to be of vital importance in demonstrating the efficacy and safety of a new therapy.

Some have also argued that randomized controls are unnecessary and unethical in studies where there are some data already available on the natural history and progression of the disease studied. Rather, they propose that a current cohort of experimentally treated patients might just as well be compared with a past cohort of patients receiving an earlier or no treatment, *i.e.*, a cohort of *historical controls.* In cases where one observes a complete dramatic reversal of the course of a disease, such as the effects of penicillin on a bacterial infection, such evidence may be convincing. However, most therapies yield modest effects and historical controls are subject to various biases that may easily skew the study results. The basic problem is that the historical control group might have very different characteristics from experimental cohort that may bias the study. Such factors might include patient selection criteria, diagnostic methods, the nature of follow-up observations, the criteria for response, and the extent of administration of concomitant medications. A difference between groups in any one of these factors or other factors could result in differences between groups with respect to study outcomes.

While most of today's scientists have embraced the randomized clinical trial, occasionally particular clinical trials arise that elicit passionate opposition on ethical grounds. A prime example is the recent clinical trials program in third-world countries on the benefits of short-term zidovudine (AZT) therapy in reducing maternal-infant HIV transmission. In a landmark clinical trial, Connor, Sperling, Gelber, *et al.* (1994) show that six weeks of AZT therapy in pregnant women with HIV reduced the transmission to the infant by two-thirds. The results of this trial were hailed in the medical community, and six weeks of antiretroviral therapy quickly became the standard of care for HIV-positive pregnant women in the United States. Unfortunately, the prohibitive cost of zidovudine has prevented developing countries from implementing what is now the standard regimen in the United States. Consequently, a large group of scientists determined that clinical trials should be conducted in these countries using a shorter, less costly regimen of antiretroviral therapy, and such trials were begun with funding from the U. S. government. In an editorial, Lurie and Wolfe

(1997) argue that placebo-controlled trials in developing countries are unacceptable, since an effective therapy had already been found in the United States:

> ... On the basis of the [Connor, Sperling, Gelber, *et al.* data], knowledge about the timing of perinatal transmission, and pharmacokinetic data, the researchers should have had every reason to believe that well-designed shorter regimens would be more effective than placebo. These findings seriously disturb the equipoise ... necessary to justify a placebo-controlled trial on ethical grounds.

In addition, they argue that, since the standard of care in developing countries (*i.e.*, not providing therapy) is not based on consideration of alternate treatments or clinical data, and rather is based on economic considerations, researchers have an ethical responsibility to provide treatment that conforms with the standard of care in the sponsoring country (*i.e.*, the U. S.).

This editorial led to much debate in the medical literature. Several of the researchers on these clinical trials in developing countries responded with their own editorial [Halsey, Sommer, Henderson, *et al.* (1997)]. They argue that a placebo control arm is necessary in order to determine if the short course of zidovudine is effective in these countries. Furthermore, they state that providing the same level of care routinely provided to mothers and their infants in the U. S. would violate the guideline to avoid undue inducements for participation in research and would make the research totally impractical.

> If these unsustainable services were provided on a temporary basis what would happen when the research project ended and local practitioners could no longer provide diagnostic tests, infant monitoring, and intensive care units necessary to support the regimen?

They close by noting that many dramatic interventions in developing countries could have been prevented had such "medical and ethical imperialism" been imposed on participants in international studies.

The Declaration of Helsinki was revised in October 2000, adding the following statement:

> The benefits, risks, burdens and effectiveness of a new method should be tested against those of the best current prophylactic, diagnostic, and therapeutic methods.

Although this does not exclude the use of placebo in studies where no proven prophylactic, diagnostic, or therapeutic method exists, this new directive is very controversial since some interpret it to mean that a placebo should never be used whenever effective therapy is available.

At issue, however, is not the act of randomization but rather the choice of the control treatment, either placebo or an active control (when the latter exists). Randomization of treatments to patients is now considered the seminal element of a clinical trial for the evaluation of a new innovation in medical care. The purpose of this book is to describe the theoretical basis for the various types or approaches to randomization commonly employed, to describe their statistical properties, and to describe considerations in their practical implementation.

1.5 PROBLEMS

1.1 From a recent issue of any major medical journal (*e.g.*, *New England Journal of Medicine*, *Journal of the American Medical Association*), select an article which presents results of a controlled clinical trial involving at least 50 patients. The study should focus on a clinical result (*i.e.*, effectiveness or safety of a treatment) rather than physiologic results (*e.g.*, laboratory or physical measurements).

(i) Give a detailed description of the study design.

(ii) Provide a critique of the study design in regard to the potential for bias in the study results or conclusions. Did the authors describe the choice of study design well and describe possible pitfalls of the design?

(iii) Based on this study, if you were the statistician for a new study (either for a new treatment for the same disease or a study confirming results of the study), describe how you would design a study using randomized controls.

(iv) Alternatively, describe how you would design a study using non-randomized controls.

(v) Discuss the implications for a randomized versus a non-randomized study on the interpretation of the results. Which would be preferable?

1.2 From a recent issue of a medical or epidemiologic journal, select an article that presents the results of a non-randomized observational study of a risk factor associated with an increase or decrease in the risk of a disease or adverse disease outcome.

(i) Give a detailed description of the study design.

(ii) Provide a critique of the study design in regard to the potential for bias in the study results or conclusions. Did the authors describe the choice of study design well and describe possible pitfalls of the design? Which possible biases are cited by the authors and what steps were taken, if any, to address them? Can you identify other possible sources of bias?

(iii) Based on this study, if you were the statistician for a new study (either for a new treatment for the same disease or a study confirming results of the study), describe how you would design a study using randomized controls, if possible.

(iv) Alternatively, describe how you would design a study using non-randomized controls.

1.3 If you were the statistician on a steering committee which is deciding whether to participate in a placebo-controlled clinical trial of maternal-infant HIV transmission and short-term AZT in a developing country, where the country has no access to the standard-of-care therapy in the United States (*i.e.*, long-term AZT therapy), what would your stance be? Prepare a five minute position paper for a classroom debate. You are asked to respond to the following questions:

(i) Are such trials necessary and ethical?

(ii) Should any placebo-controlled study be adopted?

(iii) Are studies with historical controls reasonable?
(iv) What are alternatives?

1.4 Are the considerations of individual and collective ethics the same in all clinical trials? Suppose you cross-classified a disease with respect to severity and incidence. For instance, you could have a 4-by-4 table with ordinal categories ranging from 1 to 4. For severity, the categories could range from $1 = $ mild to $4 = $ life-threatening. Similarly, incidence could range from $1 = $ very rare to $4 = $ very common. Within each cell of the cross-classification, determine the relative importance of individual versus collective ethics. (Palmer and Rosenberger, 1999)

1.6 REFERENCES

BRESLOW, N. E. AND DAY, N. E. (1980). *Statistical Methods in Cancer Research. Volume I - The Analysis of Case-Control Studies.* International Agency for Research on Cancer, Lyon.

BRESLOW, N. E. AND DAY, N. E. (1987). *Statistical Methods in Cancer Research. Volume II - The Design and Analysis of Cohort Studies.* International Agency for Research on Cancer, Lyon.

BUYSE, M. E., STAQUET, M. J., AND SYLVESTER, R. J. (1984). *Cancer Clinical Trials: Methods and Practice.* Oxford Medical Publications, New York.

CAIRNS, J. (1985). The treatment of diseases and the war against cancer. *Scientific American* **253:5** 51–60.

CHOW, S-C. AND LIU, J-P. (1998). *Design and Analysis of Clinical Trials.* Wiley, New York.

CONNOR, E. M., SPERLING, R. S., GELBER, R., KISELEV, P., SCOTT, G., O'SULLIVAN, M. J., VANDYKE, R., BEY, M., SHEARER, W., JACOBSEN, R. L., JIMINEZ, E., O'NEILL, E., BAZIN, B., DELFRAISSY, J-F., CULNANE, M., COOMBS, R., ELKINS, M., MOYE, J., STRATTON, P., BALSLEY, J., FOR THE PEDIATRIC AIDS CLINICAL TRIALS GROUP PROTOCOL 076 STUDY GROUP. (1994). Reduction of maternal-infant transmission of human immunodeficiency virus type 1 with zidovudine treatment. *New England Journal of Medicine* **331** 1173–1184.

CORNFIELD, J. (1959). Principles of research. *American Journal of Mental Deficiency* **64** 240–252.

DIABETES CONTROL AND COMPLICATIONS TRIAL RESEARCH GROUP. (1993). The effect of intensive treatment of diabetes on the development and progression of long-term complications in insulin-dependent diabetes mellitus. *New England Journal of Medicine* **329** 977–986.

ENGELHARDT, H. T. (1996). *The Foundations of Bioethics.* Oxford University Press, New York.

FISHER, R. A. (1935). *The Design of Experiments.* Oliver and Boyd, Edinburgh.

FRANCIS, T. (1955). An evaluation of the 1954 poliomyelitis vaccine trials – summary report. *American Journal of Public Health* **45** 1–63.

FREEDMAN, D., PISANI, R., PURVES, R. (1998). *Statistics.* Norton, New York.

FRIEDMAN, L. M., FURBERG, C. D., AND DEMETS, D. L. (1998). *Fundamentals of Clinical Trials.* Springer, New York.

HALSEY, N. A., SOMMER, A., HENDERSON, D. A., AND BLACK, R. E. (1997). Ethics and international research. *British Journal of Medicine* **315** 965.

KELSEY, J. L., WHITTEMORE, A. S., EVANS, A. S., AND THOMPSON, W. (1996). *Methods in Observational Epidemiology.* Oxford University Press, New York.

LEWIS, E. J., HUNSICKER, L. G., LAN, S., ROHDE, R. D., LACHIN, J. M., AND THE LUPUS NEPHRITIS COLLABORATIVE STUDY GROUP. (1992). A controlled trial of plasmapheresis therapy in severe lupus nephritis. *New England Journal of Medicine* **326** 1373–1379.

LURIE, P. AND WOLFE, S. M. (1997). Unethical trials of interventions to reduce perinatal transmission of the human immunodeficiency virus in developing countries. *New England Journal of Medicine* **337** 853–856.

MATTHEWS, J. N. S. (2000). *An Introduction to Randomized Controlled Clinical Trials.* Arnold, London.

PALMER, C. R. AND ROSENBERGER, W. F. (1999). Ethics and practice: alternative designs for randomized phase III clinical trials. *Controlled Clinical Trials* **20** 172–186.

PIANTADOSI, S. (1997). *Clinical Trials.* Wiley, New York.

POCOCK, S. J. (1984). *Clinical Trials: A Practical Approach.* Wiley, New York.

ROSENBAUM, P. R. (1995). *Observational Studies.* Springer, New York.

SELVIN, S. (1996). *Statistical Analysis of Epidemiologic Data.* Oxford University Press, New York.

SURGEON GENERAL. (1982). *The Health Consequences of Smoking: Cancer.* U. S. Department of Health and Human Services, Public Health Service, Rockville.

TYGSTRUP, N., LACHIN, J. M., AND JUHL, E. (1982). *The Randomized Clinical Trial and Therapeutic Decisions.* Marcel Dekker, New York.

2

Issues in the Design of Clinical Trials

2.1 INTRODUCTION

Whereas laboratory science is performed in a carefully controlled and monitored environment, clinical trials are experiments which are conducted in the workplace of medicine: physician's offices, clinics, or hospitals, as opposed to laboratories. Many clinical trials are *multi-center*, that is, they are performed by a group of participating clinics or care units, and hence are conducted by a large network of nurses, research coordinators, and physicians. Large amounts of data on study subjects are recorded and computerized. To complicate matters even further, study subjects are human beings, who are often asked to self-administer study treatments at home. So clinical trials are a complex, collaborative effort involving physicians, nurses, computer scientists, data managers, and statisticians. And guiding everyone in this collaborative effort is the *study protocol*, a document that describes the aims, procedures, and official policies of the scientific endeavor. The importance of the protocol cannot be understated: a study participant who violates protocol may bias the study and make any conclusions invalid. In this chapter, we will discuss design issues in clinical trials that every protocol should address.

2.2 STUDY OUTCOMES

The ultimate basis for any scientific investigation is the statement of its objectives. For a clinical trial, the specific aims should be stated so as to define the target population, the time course of observation and, perhaps most important to the statistician, the outcome measures. Such polemics are easily stated, but unfortunately are difficult to implement. This stage is crucial, however, because all other design features stem from the statement of objectives, including the statistical analysis plan.

In general, short-term, fixed-duration clinical trials tend to be focused on direct estimation of a treatment effect in terms of the difference of means, rates, or proportions, for example, between the group assigned to the experimental innovation (the treated group) versus the control group. Longer-term variable follow-up trials are often focused on the time to some event, such as death, or some measure of disease progression. Such trials are called *survival trials*, a generic term that is used to describe time-to-event outcomes, where the event need not be death, such as time to disease progression, or to remission or even to healing in some cases.

In virtually any disease there are defined stages of worsening (or improvement) of severity which are used in clinical research, if not in clinical practice, to describe the stages of disease progression. To the extent possible, the objectives of the clinical trial, and hence its outcome measures, should be defined in terms of clinically relevant indices of disease progression, or *clinical effectiveness.*

The statistician's responsibility is to help the investigators to frame the statement of objectives in such a manner that a clinically relevant primary outcome measure and a testable statistical hypothesis are specified, from which the primary statistical analysis is also specified. This primary outcome analysis will drive the design of the study: its length, the number of subjects to be randomized (see Section 2.6), and the statistical analysis plan. Leading a group of investigators to a single primary hypothesis may be one of the greatest challenges that a statistician ever faces. It should be noted that clinical trials are often large enough to answer many other interesting secondary hypotheses, including the effects on secondary outcome measures, or the effectiveness of treatments within subgroups. However, the design should be impelled by a single primary outcome measure of clinical effectiveness.

It is tempting, but dangerous, to plan a clinical trial to only elicit information on the *biological activity* of a therapy (*e.g.*, the effect of a drug on tumor size in cancer or CD4 levels in AIDS). Such information can be elicited quickly and easily. But biological activity is only a *surrogate outcome* for a meaningful outcome of interest in a clinical trial that reflects clinical effectiveness. Clinical effectiveness unequivocally affects patients in a tangible way; for example, by lengthening life (survival time) or increasing quality of life. These outcomes take much longer to ascertain, but a valid clinical trial should be able to determine the true clinical outcome of patients on a therapy.

Prentice (1989) proposed a set of statistical criteria that should be satisfied for concluding that a treatment would favorably affect a clinical outcome based on demonstration of a treatment effect on a biological surrogate outcome. Based on these and other considerations, Fleming and DeMets (1996) present four models in which a surrogate outcome is inappropriate in determining clinical effectiveness:

Model 1. The disease affects the surrogate and the true clinical outcome, but independently. For example, smoking causes yellow fingers and causes lung cancer and death, but an intervention that reverses yellow fingers (the surrogate outcome) may do nothing to reduce premature deaths due to smoking (clinical effectiveness).

Model 2. The disease affects the true outcome via the surrogate, and the intervention bypasses the surrogate. For example, a drug may indeed improve survival (clinical effectiveness) but not have any impact on CD4 counts in AIDS (surrogate).

Model 3. The disease affects the true outcome via a surrogate, but an intervention targeting the surrogate outcome causes adverse effects with respect to the clinical outcome. The literature is replete on drugs that have an effect on biological activity, but have been shown to have no effect or a deleterious effect on survival or other clinical outcome. For example, encainide and flecainide reduced arrhythmias, but relative to placebo, tripled the death rate (Echt, Liebson, Mitchell, *et al.*, 1991).

Model 4. The disease affects the true outcome via a surrogate, but intervention targeting the clinical outcome has no effect on the surrogate. For example, gamma interferon contributed to a 70 percent reduction, relative to placebo, in infections in children with chronic granulomatous disease, yet had no effect on killing bacteria (International Chronic Granulomatous Disease Cooperative Study Group, 1991).

Fleming and DeMets give two criteria for evaluating the relevance of a surrogate outcome in a clinical trial. First, it must be correlated with the clinical outcome. Second, it must fully capture the net effect of the treatment on the clinical outcome. The second criterion is often difficult to determine. Validating a surrogate requires a comprehensive understanding of causal path of disease process and the intervention's intended and unintended effect. Therefore, measures of biological activity should be used with caution as outcomes in clinical trials.

Nevertheless, some definitive outcomes in chronic diseases, such as death, may occur so far in the future that the clinical outcome is not logistically ascertainable. As an example, consider the study of captopril, an angiotensin converting enzyme (ACE) inhibitor, in progressive diabetic nephropathy or kidney disease, one of the complications of diabetes mellitus. The earliest stage of nephropathy is the leakage of tiny amounts of albumin into urine. When the leakage reaches the level that can be detected using an ordinary "dip stick" in urine, about 300 mg/24 hours, the subject has developed overt proteinuria at which point nephropathy is well established. The process will ultimately lead to total destruction of all of the functioning glomeruli that are the biological filters in the kidney and the patient enters renal failure. Life can then be sustained either by dialysis or a renal transplant. Animal studies showed that captopril might reduce the rate of progression to renal failure among patients with proteinuria. This process, however, could take many years, and a clinical trial designed to demonstrate an effect on the incidence of renal failure was considered unfeasible.

In clinical practice, the concentration of creatinine in serum (mg/dL) is universally employed as a simple measure of renal function and to monitor the decline in renal function over time. When the glomerlular filtration rate (GFR) falls below the normal range the level of serum creatinine begins to rise. Although it might be tempting to perform a fixed duration trial to compare the mean rise in serum creatinine from baseline between the treatment groups, such an analysis would deal only with group means, rather than individual response to therapy. A more appropriate design would be to employ an outcome that is the time to a specific "event" of clinical relevance in individual patients. Such a design would provide a better description of clinical progression in the population than would a simple comparison of means. Further, since each patient is followed to the time of an outcome that represents clinical progression in that individual, consequently the outcome of the trial represents the treatment effect on the incidence of a clinically relevant outcome.

Earlier studies had shown that the inverse creatinine declined linearly over time in patients with established nephropathy (proteinuria). Thus the study was designed to detect a treatment effect on the time to doubling of the baseline serum creatinine, or the time to a 50 percent reduction in the GFR. While no studies were available to show that the Prentice or Fleming-DeMets criteria were satisfied, virtually all physicians would agree that a treatment effect on this outcome is highly meaningful. The trial demonstrated a 48 percent reduction ($p < 0.007$) in the incidence (hazard) of renal progression using this outcome (Lewis, Hunsicker, Bain, *et al.*, 1993). Despite the smaller number of events, a 50 percent reduction was observed in the risk of death or renal transplant ($p < 0.006$).

2.3 SOURCES OF BIAS

The objective of any clinical trial is to provide an unbiased comparison of the differences between two treatments. As we shall see in the next chapter, the randomization of subjects between the treatment groups is the paramount statistical element that allows one to claim that a study is unbiased. However, randomization alone does not provide an unbiased study. As Lachin (2000) points out, randomization is necessary, but alone is not sufficient. Two other requirements are (i) the outcome assessments should be obtained in a like and unbiased manner for all patients; and (ii) data that are missing, if any, from randomized patients do not bias the comparison of the treatment groups. Point (i) reflects the importance of standardization and masking and point (ii) the importance of the statistical analysis philosophy

2.3.1 Standardization and masking

All clinical trials should employ a standard system of outcome evaluations in all patients randomized. The objective is to ensure that all subjects are evaluated in an unbiased and precise manner regardless of treatment assignment and the response to treatment. This is most readily achieved by employing a uniform schedule of outcome assessments for all patients with a single central unit for the evaluation of the outcome evaluations in all patients.

To the extent possible, all trials should also be *double-masked*, meaning that neither the patient nor physician are aware of the treatment randomly assigned to the patient. Under no circumstance should the masking be broken, unless there is a serious adverse event that requires knowledge of the assigned treatment. Although it is clear that the patient should not know the treatment assignment, it is often questioned why the physician or care-giver should not be informed. If the physician knows or can guess what treatment will be assigned next, he or she could bias the study by selecting a patient more likely to benefit from the treatment. The physician may also treat patients differently according to which treatment the patient is taking. In these ways, subtle biases can influence the results. These biases can be mitigated by double-masking.

Complete double-masking may not be possible in some clinical trials, for example, clinical trials of surgical procedures where the surgeon must know which procedure to perform. In this case, it is then preferable that outcome evaluations be masked to treatment assignment to the extent possible, where the evaluator is unaware of the treatment assignment. For example, in a clinical trial of laser therapy for glaucoma, the outcome might be a deterioration in the visual field. The visual field tests, however, could then be forwarded to a central reading facility where the readers are masked to the treatment assignments of individual eyes. Another example of a clinical trial in which double-masking was impossible is the Diabetes Control and Complications Trial, where patients were randomly assigned to receive either conventional or intensive blood glucose control management. However, investigators who evaluated principal outcome measures were masked to treatment assignments. While complete masking of treatment may not be possible, the evaluation of outcomes can almost always be masked. This is one reason that many trials employ a central laboratory or reading center. Another reason is that it is easier to control the accuracy and precision of measurements (*i.e.*, quality control) of a central laboratory than those of multiple laboratories.

In such trials where the randomization is unmasked to the recruiting physician or care-giver, if the randomization procedure is predictable, it is possible for the physician or care-giver to bias the composition of the treatment groups by attempting to predict the next assignment and choosing a patient the physician would prefer to receive that treatment. In Chapter 6, we examine a model for *selection bias* (Blackwell and Hodges, 1957) which quantifies the bias resulting from an experimenter's trying to guess the treatment assignments. There we show that different randomization procedures have different susceptibility to such bias. Further, if treatment assignments are generated by a simple independent Bernoulli sequence (*i.e.*, coin-tossing), then it is impossible to predict the treatment to be assigned and the selection bias potential is zero.

Another related consideration is the implementation of the study treatment regimens and the clinical management of patients under follow-up. By definition, a clinical trial entails the treatment of patients (or healthy individuals) under scientific conditions. In order for the results to have an impact on the practice of medicine, the treatment procedures employed must be precisely described. For the results to be scientifically rigorous, all aspects of the treatment and clinical management of patients should be standardized as much as possible. These considerations are especially important in a multi-center trial in which it is important that each clinical team ideally should treat and manage patients in an identical manner. This is also important statistically. Virtually everything that happens to a patient after randomization into the trial is a potential outcome measure, especially clinical events which reflect progression of a patient's disease, or events which reflect an adverse effect of treatment. For these reasons, to the extent possible, the trial should define and standardize all aspects related to the administration of the study treatments, the ascertainment of clinical events, and clinical management.

2.3.2 Statistical analysis philosophy

There are two prevailing philosophies in the analysis of clinical trials, especially in studies of pharmaceuticals (drugs). On one side is the pharmacologist who wishes to assess the pharmacologic efficacy of the regimen. In this sense, an *efficacy analysis* is performed using the subset of patients who are able to tolerate the drug, are adequately compliant, and to whom the agent is effectively administered. The basic strategy is to examine the experience of the patients entered into the trial, and then to select the subset of these patients that meet the desired efficacy criteria for inclusion into the analysis. On the other side is the clinician or regulatory scientist who wishes to assess the overall clinical effectiveness, meaning the outcomes of all patients for whom the treatment is initially prescribed, irrespective of potential side effects or incomplete administration. Although compliance is an important determinant of ultimate effectiveness, the therapeutic question is to assess the effectiveness of the treatment in a population of ordinary subjects with variable degrees of compliance. Such an analysis is called *intention to treat*, because the outcome is compared between two samples that are initially assigned to receive different treatments, regardless of the level of tolerance or compliance. Such an analysis attempts to assess the long-term effects in the population of an initial treatment decision to adopt one regimen versus another. In order to conduct an intention to treat analysis, therefore, all subjects randomized into the study must be evaluated as scheduled during follow-up, regardless of the extent of compliance with the treatment protocol or the occurrence of adverse effects.

Following Lachin (2000), is easy to see how bias might enter the study under an efficacy analysis. If one starts a study with 100 patients who are randomized equally between two treatment groups, but at the end of the study outcome assessments are obtained in only 60 of these, then those 60 patients may not be unbiased. This is because the observations missing for the 40 patients may not be *missing completely at random (MCAR)*, meaning that the presence or absence of an observation occurs purely by chance. For example, a patient on placebo may choose to be noncompliant because he or she feels the treatment is not effective. Likewise, a patient who begins to feel better on an experimental therapy may opt to discontinue medication during the course of the study. In these cases, missingness depends on the outcome of interest, and the remaining subset analysis ignores important information on effectiveness of the treatment in missing patients. Consequently, the only incontrovertibly unbiased study is one in which all randomized patients are evaluated and included in the analysis, and this is the essence of the intent-to-treat philosophy. It should be very clear that final outcome ascertainment should be the investigator's goal for each individual patient enrolled in the study, regardless of their level of active participation in the trial. Thus the essence of the intention to treat design is to ensure that every patient randomized is followed and evaluated as scheduled until either death or the end of the trial.

Thus the core tenet of the intent-to-treat principle is an intent-to-treat design in which all subjects randomized are followed as specified under the original protocol, regardless of compliance with the treatment regimens, or adverse effects, or whatever; the only exceptions being death, a clinical proscription against the follow-up procedure, or withdrawal of patient consent.

2.3.3 Losses to follow-up and noncompliance

Various authors have used the terms "losses to follow-up," "dropouts," and "noncompliance" interchangeably. In this book, we use the term *lost to follow-up* to describe patients who do not continue follow-up visits for primary outcome assessments. Such patients may have moved away or may no longer be willing to participate in the study for various reasons. Provided that the reason lost to follow-up is not related to the outcome of the study, the data on these patients are MCAR, and should not bias the study. Every effort should be made to ascertain the reason these patients dropped out of the study, and if it is at all treatment-related. Adjustments to the number of patients randomized are usually built into the study to accommodate a small number of losses to follow-up, as we will see in Section 2.6.4. By *noncompliance*, we refer to less than maximally effective treatment in a patient who continues follow-up. These patients should be included in an intention to treat analysis, to avoid bias.

Even when placebo controls are used to implement double-masking, some drugs, for example, have a known adverse effect profile, such as where the drug is known to induce mild hepatotoxicity, or gastrointestinal disturbances, etc. In such cases, substantial biases may be introduced if subjects who experience such adverse effects are terminated from further follow-up. In such cases the resulting missing observations are not missing at random and it is not possible to argue that the resulting observed measures are unbiased.

2.3.4 Covariates

When comparing two groups in the context of a medical therapy, it is critical that the two groups be comparable with respect to important covariates. These covariates may be known in advance or unknown, and treatment imbalances with respect to these covariates can bias the study. When there are known covariates, then strategies can be used to force the randomization procedure to balance the distribution of covariates among groups, or to promote balance. These techniques will be discussed in Chapter 4. However, it is still possible that imbalances may occur with unknown covariates, which would lead to what is commonly referred to as *accidental bias*. Randomization can mitigate accidental bias, and the randomization procedures can be distinguished, at least theoretically, by the susceptibility to such accidental covariate imbalances, or accidental bias. This is discussed in Chapter 5.

In addition to assessing or describing the "balance" of the randomization with respect to known covariates, measured covariates are also used to assess the association of covariate values with the outcome of the study, and also to assess the treatment group effect as a function of covariate values, such as separately between men and women; we call these *treatment by covariate interactions*. Randomization has no impact on whether treatment by covariate interactions will exist since these are true characteristics of the phenomena under study, not the result of chance.

The Diabetes Prevention Program (Diabetes Prevention Program Research Group, 2002) showed that treatment with the drug metformin versus placebo provides a 31 percent reduction ($p < 0.001$) in the risk of developing type 2 diabetes in individuals with impaired glucose intolerance. Among the important analyses was an assessment of this treatment effect among subjects stratified into subgroups defined by the baseline level of body mass index (BMI) in kg/cm^2. Among those with BMI < 30, metformin provided only a 3 percent risk reduction versus placebo, whereas the risk reduction was 16 percent among those with $30 \leq BMI < 35$, and 53 percent among those with $BMI \geq 35$. The heterogeneity of treatment effect among these BMI subgroups was significant at $p < 0.05$. Thus, balancing treatment groups on the level of BMI would not alone lead to correct conclusions: that the drug was effective only in a certain subgroup of patients and not in another subgroup. If such a subgroup analysis is known to be important in advance, studies can be powered accordingly to detect a treatment by covariate interaction, but in practice, studies often do not have enough power to detect such interactions, and such subgroups may not be known in advance. Consequently, it is important to remember that, while randomization tends to induce independence between the treatment effect and unobserved covariates by eliminating accidental bias, it does not eliminate interactions.

2.4 EXPERIMENTAL DESIGN

The essence of a clinical trial is the comparison of the effects of the experimental treatment to those of a control treatment. For example, in a therapeutic trial to compare a stated dose of a drug versus a placebo, if the other aspects of the trial are rigorous with regard to controlling bias, then a sharp comparison can be made to discern the clinical effects of the drug. Although there may be multiple factors that contribute to the effectiveness of an experimental treatment, trials are not usually designed to elucidate those precise factors, but only to determine if the treatment is effective. In most trials, therefore, the experimental design is quite simple. Usually only two or more groups are employed to compare two or more treatments. The treatment effect can, for example, be estimated from a simple one-way group comparison.

While *factorial designs* [*cf.* Cochran and Cox (1957)] are frequently employed in other sciences, they are rarely used in clinical trials. The exception is the consideration of combination therapies, where a 2-way factorial design may be employed. An analysis of variance strategy is then employed, where treatment A and treatment B are main effects and there is an interaction term for treatments A and B. The test of interaction assesses whether the combination has effects above and beyond each treatment individually. However, such studies are often designed with the assumption that there is no interaction, and are then underpowered for a test of interaction. If an interaction is later observed, then the effects of each level of A depend on the level of B and vice versa. In this case, the study also is underpowered for the detection of nested effects of factor A within each level of B and vice versa. In addition, if a binary or time to event outcome analysis is employed, even when there is no interaction but a marginal treatment effect on one factor is present, the power for the assessment of the treatment effect on the other factor will be reduced. Rarely are higher order factorial designs employed.

Short-term clinical trials of chronic, but stable, conditions will sometimes employ a *crossover design* [*cf.* Jones and Kenward (1989)], whereby each patient will receive one of two treatments initially, and then, after a *washout period* of withdrawal from the treatment, the patient receives the other treatment. For the analysis of this design, the results of treatment A are compared to those of treatment B in aggregate over all patients. Thus, the marginal statistic for treatment A actually consists of the effect of treatment A during the first period of administration plus the residual effect of treatment A during the second period of administration (after having received treatment B during the first period). So one must assume that there are no residual carryover effects from the first period of administration to the second period of administration. The advantage of crossover trials is that they employ half as many patients, since each patient serves as his or her own control. However, the condition of no carryover effect is difficult to satisfy, and consequently they are rarely used in long-term clinical trials which are designed to assess the effects of treatment on the clinical course of a disease.

In conclusion, the most widely employed design for clinical trials is the simple two-group comparison design. Therefore, it is in this context that this book will address the randomization of subjects to receive one of the study treatments.

2.5 RECRUITMENT AND FOLLOW-UP

In a typical clinical trial, patients are identified for screening and consideration for entry into the trial over a period of time, often years. The interval is called the *recruitment period* during which patients are screened, and if found eligible, are then randomized to one of the study treatments. Eligibility requirements are agreed upon by the investigators before the trial begins and are recorded in the study protocol. Eligibility requirements are designed to ensure that a homogeneous strain of patients is recruited for the trial. If eligibility criteria are changed or relaxed at any point in the trial, shifts in patient characteristics may occur, which can cause problems in the statistical analysis. Likewise, if protocol violations occur where an investigator randomizes an ineligible patient, serious biases may result. The process of recruitment of patients over an interval of time results in what is termed *staggered entry*, because all patients do not enter the trial simultaneously at a given point in calendar time. The clinical trial then systematically collects observations over time according to a follow-up schedule. The follow-up schedule specifies the duration of treatment of each patient and the precise times during follow-up at which specific procedures are performed and measurements obtained.

The two common plans for the design of clinical trials are either *fixed* or *variable* follow-up duration. Each provides for a period of recruitment in calendar time of length R, where time 0 is the calendar date at which the first patient enters the trial and time R is the subsequent date on which the last patient is randomized. The trial is then continued until calendar time $T, T > R$, which provides for the follow-up of the last patient entered. The difference between the time a subject enters a trial and follow-up is completed is the *study time* of that subject. In a fixed follow-up duration trial, each patient is followed for the same prescribed period of time, regardless of when that patient was entered into the study. Thus, each patient's study time is the same, and the study cannot be concluded until each patient completes that study time. Fixed duration trials are often employed to assess short-term objectives, such as when each patient is treated with a drug for a two-month interval. Many long-term studies employ variable follow-up duration, where patients are followed until a common closing date, regardless of when they were randomized. Thus, a patient's study time depends on the date the patient entered the trial. For example, in a five-year trial ($T = 5$) with a two-year recruitment period ($R = 2$), the first patient entered would be followed for all $T = 5$ years, whereas the last patient entered would be followed for $T - R = 3$ years. Assuming recruitment follows a uniform distribution, average duration of follow-up would be 4 years with a standard deviation of 0.57 years.

Uniform recruitment assumes the distribution function G of patient entry times is linear over $[0, R]$. It is not unusual, though, for G to be convex or concave. Convexity implies that recruitment is initially faster than expected under uniform entry and then declines; concavity implies that recruitment is initially slower than expected and then increases. One possible model is the truncated exponential distribution, where we assume patient entry times $Z_1, ..., Z_n$ are independent and identically distributed with density function

$$g(z) = \frac{\gamma e^{-\gamma z}}{1 - e^{-\gamma R}}, 0 \leq z \leq R, \gamma \neq 0. \tag{2.1}$$

If $\gamma > 0$, G is convex, and if $\gamma < 0$, G is concave. Under model (2.1), average duration of follow-up is given by

$$T - \frac{1 - e^{-\gamma R} - \gamma R e^{-\gamma R}}{\gamma(1 - e^{-\gamma R})}, \tag{2.2}$$

with a standard deviation of

$$\frac{\left(1 - e^{-2\gamma R} - (2 + \gamma^2 R^2)e^{-\gamma R}\right)^{1/2}}{\gamma(1 - e^{-\gamma R})}. \tag{2.3}$$

In our example with $R = 2$ and $T = 5$, if $\gamma = 1$, average duration of follow-up is 4.31 years with a standard deviation of 0.53 years. If $\gamma = -1$, average duration of follow-up is 3.69 years with a standard deviation of 0.53 years (note the symmetry).

With either fixed duration or variable follow-up, the duration of follow-up should be based on the period of time needed to assess the trial's objectives. The frequency of follow-up visits is usually based on conventional clinical practice for the treatment and follow-up of the condition under study. However, in other instances, the frequency of assessment may be based on other considerations, such as the frequency required to chart the incidence of an event or a change in a characteristic over time, or to safeguard patient safety by toxicity screens. During these visits, ascertainment of the study outcome, medical assessments to determine safety of the therapy, determination of compliance with the study's treatment regimen, and any other medical procedures dictated by the standard-of-care for the condition will be performed. In so far as possible, clinical trials should mimic usual clinical practice.

2.6 DETERMINING THE NUMBER OF RANDOMIZED SUBJECTS

In the planning stages of a randomized clinical trial, it is necessary to determine the numbers of subjects to be randomized. While the exact final number that contribute to any analysis will be unknown, due to losses to follow-up and staggered entry, it is still desirable to determine a target sample size based on some model. This *sample size estimate* will then allow estimates of the total cost of the trial, the number of clinics required, and target recruitment numbers, etc. Typically, the number of subjects is computed to provide a fixed level of power under a specified alternative hypothesis [see, for example, Lachin (1981) and Donner (1984)]. The alternative hypothesis usually represents a minimal, clinically-meaningful treatment effect. Power (1− probability of a type II error) is an important consideration for several reasons. Low power can cause a truly beneficial therapy to be rejected. However, too much power may make results statistically significant that are not clinically significant. Standard regulatory criteria for clinical trials often lead to specifying the probability of type I error (α) to be 0.05 and power to be 0.80 to 0.90. However, such specifications belie the consideration of the relative cost of a type I or type II error in a particular study. There are examples of studies where investigators determined that a type II error was so much more costly than a type I error, that α was fixed far from 0.05 [see, for example, Samuel-Cahn and Wax (1986)].

Tests of the treatment effect in clinical trials are typically two-sided, for two principal reasons: first, it is usually relevant if the placebo or standard therapy is more efficacious than the experimental therapy; and, second, even if only a one-sided hypothesis is really of interest, a two-sided test requires a more stringent 0.025-level test which gives added protection from a type I error.

2.6.1 Development of the main formula

Under this framework of power considerations, it is necessary to assume a *population model*. Let n be the total number of subjects randomized in the trial and let n_i be the number randomized to treatment group i. For two treatments $i = A, B$, say, $n = n_A + n_B$. We assume here that the allocation proportions are known in advance,

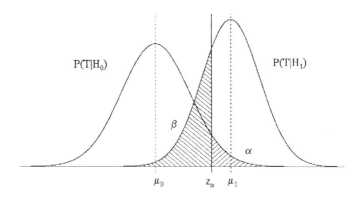

$P(T|H_0)$ $P(T|H_1)$

β

α

μ_0 z_α μ_1

Fig. 2.1 *Distribution of a test statistic T under the null and alternative hypotheses, with the rejection region of size α and a type II error rate of size β. (Lachin (2000, p. 64); reprinted with permission of John Wiley and Sons.)*

i.e., that $Q = n_A/n$ and $1 - Q = n_B/n$ are predetermined. This will not be the case under adaptive randomization, as we will see later in the book. Under a population model, it is assumed that responses $Y_{ij}, j = 1, ..., n_i$, are independently and identically distributed according to some known distribution $G(y_{ij}|\theta_i)$, where θ_i is possibly a vector-valued parameter associated with the ith population. For example, if θ_i is a single parameter representing an outcome associated with treatment, a standard hypothesis test would be $H_0 : \theta_A = \theta_B$ versus $H_1 : \theta_A \neq \theta_B$. Let S_n be a statistic to test a hypothesis regarding the equality of one or more members of θ. Based on the distribution of the measurements, $Y_{ij} \sim G(y_{ij}|\theta_i)$, it is usually easy to derive the distribution of a statistic S_n under H_0 and H_1. The central limit theorem will usually lead to a normal distribution under H_0 and H_1, such as the following:

$$H_0 : S_n \sim N(\mu_0, \Sigma_0^2), \; H_1 : S_n \sim N(\mu_1, \Sigma_1^2), \tag{2.4}$$

where μ_0 and μ_1 are functions of θ_i and Σ_0 and Σ_1 are functions of n and θ_i. This provides a large sample test of H_0 of the form $T_n = (S_n - \mu_0)/\Sigma_0$, which is asymptotically distributed as standard normal under H_0. With this test, H_0 is rejected at level α if $|T_n| \geq z_{\alpha/2}$ (two-sided) or $T_n \geq z_\alpha$ (one-sided), where z_α is the standard normal deviate; *i.e.*, if Z is a standard normal variate, $\Pr(Z \geq z_\alpha) = \alpha$. The power of the test Z is given by $1 - \beta$. Figure 2.1 shows the relationship between α and β for the standard hypothesis testing problem.

The basic relationship used to derive n, based on values of β and α under a specified alternative hypothesis can be derived as follows. Under the distributional assumption in (2.4), for a one-sided alternative ($\mu_1 > \mu_0$),

$$
\begin{aligned}
1 - \beta &= \Pr\left(T_n > z_\alpha \middle| H_1\right) \\
&= \Pr\left(\frac{S_n - \mu_0}{\Sigma_0} > z_\alpha \middle| H_1\right) \\
&= \Pr(S_n > \mu_0 + z_\alpha\Sigma_0 | H_1) \\
&= \Pr\left(\frac{S_n - \mu_1}{\Sigma_1} > \frac{\mu_0 + z_\alpha\Sigma_0 - \mu_1}{\Sigma_1} \middle| \frac{S_n - \mu_1}{\Sigma_1} \sim N(0,1)\right) \\
&= \Pr\left(Z > \frac{\mu_0 - \mu_1}{\Sigma_1} + z_\alpha\frac{\Sigma_0}{\Sigma_1}\right),
\end{aligned}
$$

where Z is a standard normal variate. This implies that that

$$
-z_\beta = \frac{\mu_0 - \mu_1}{\Sigma_1} + z_\alpha\frac{\Sigma_0}{\Sigma_1}.
$$

Simple algebra then leads to the equation

$$
\mu_1 - \mu_0 = z_\alpha\Sigma_0 + z_\beta\Sigma_1. \tag{2.5}
$$

For a two-sided test, (2.5) is given by

$$
|\mu_1 - \mu_0| \cong z_{\alpha/2}\Sigma_0 + z_\beta\Sigma_1. \tag{2.6}
$$

For more than two groups, formulas can be adjusted accordingly, and a Bonferroni correction is standardly used for more than two hypothesis tests. For example, in the NIH-sponsored benign prostatic hyperplasia clinical trial, patients were randomized to one of four groups: placebo, finasteride, doxazosin, or finasteride and doxazosin. Each of the active therapy groups was to be compared to the placebo group, for a total of three hypothesis tests. Hence, for a two-sided test, a Bonferroni adjustment led to the term $z_{\alpha/6}$ in equation (2.6).

2.6.2 Example

Consider a comparison of two means. Here we assume $Y_{ij} \sim N(\nu_i, \sigma_i^2), i = A, B, j = 1, ..., n_i, n_A = Qn$. Assume $\sigma_A^2 = \sigma_B^2 = \sigma^2$, and σ^2 is known. Then $\theta_i = (\nu_i, \sigma^2)$. We wish to test $H_0 : \nu_A = \nu_B$ versus $H_A : \nu_A \neq \nu_B$, so that $\mu_0 = 0$ and $\mu_1 = \nu_A - \nu_B$, using the statistic $S_n = \bar{y}_A - \bar{y}_B$, where $\bar{y}_i = \sum_{j=1}^{n_i} y_{ij}/n_i$. Then $\Sigma_0^2 = \Sigma_1^2 = [Q(1-Q)]^{-1}\sigma^2/n$, and equation (2.6) becomes

$$
|\mu_1| = (z_{\alpha/2} + z_\beta)\sigma/\sqrt{Q(1-Q)n} \tag{2.7}
$$

or

$$
n = \frac{1}{Q(1-Q)}\left(\frac{(z_\alpha + z_\beta)\sigma}{\mu_1}\right)^2.
$$

Similar expressions can be derived for differences of binomial probabilities or Poisson rates (Problems 2.4 and 2.5).

This example presents the statistician with two dilemmas: σ^2 is assumed to be known, which is always an unreasonable assumption, and we can only compute the required number of subjects under a specified alternative. For both problems, the statistician must rely on the physician who is familiar with the natural history of the disease under no treatment (in the case of a placebo-controlled trial) or under the standard therapy (in a trial comparing a new and existing therapy). The physician must provide an estimate of σ^2 using information from previous studies or from his or her experience. The particular alternative specified should represent the minimal clinically significant difference for which the physician would declare the experimental therapy to be successful or worthwhile in practice.

2.6.3 Survival trials

In a survival trial where the study objective is to compare a time-to-event outcome between two treatment groups, equation (2.6) can be used to derive the number of randomized subjects under a parametric failure-time model. Lachin and Foulkes (1986) assume an exponential model for testing the equality of two hazard functions, λ_A and λ_B. Since the logrank test for equality of hazard (survival) functions over time is known to be the asymptotically most efficient test under a proportional hazards model, and since the exponential distribution is the special case of proportional constant hazards, the exponential model leads to estimates of the sample size and power function of the logrank test. If the actual hazard in the control group fluctuates over time, calculations based on the exponential model will still be adequate provided that the expected number of events in the control group under the model agrees with the actual expected value. If not, then alternate methods, such as those of Lakatos (1988) are preferred.

For the test $H_0 : \lambda_A = \lambda_B$ versus $H_A : \lambda_A \neq \lambda_B$, $S_n = \hat{\lambda}_A - \hat{\lambda}_B$, the maximum likelihood estimator of $\mu_1 = \lambda_A - \lambda_B$. Let $\bar{\lambda} = Q\lambda_A + (1 - Q)\lambda_B$. Under H_0, the asymptotic distribution is $N(0, \Sigma_0^2)$, and under H_1, the asymptotic distribution is $N(\mu_1, \Sigma_1^2)$, where $\Sigma_0^2 = \phi(\bar{\lambda})/nQ(1-Q)$ and $\Sigma_1^2 = ((1-Q)\phi(\lambda_A) + Q\phi(\lambda_B))/nQ(1 - Q)$. The function ϕ will depend on the patient entry distribution and losses to follow-up. Substituting into equation (2.6), we obtain

$$n = \frac{\left\{ z_{\alpha/2} \left(\phi(\bar{\lambda}) \right)^{1/2} + z_\beta \left((1 - Q)\phi(\lambda_A) + Q\phi(\lambda_B) \right)^{1/2} \right\}^2}{\mu_1^2 Q(1 - Q)}. \tag{2.8}$$

One would use z_α in place of $z_{\alpha/2}$ for a one-sided test. Note that this equation requires specification of λ_A and λ_B, which can be estimated if one knows something about the cumulative incidence functions on treatments A and B. The hazard function can be computed from the incidence rate over T years, ρ, by the formula $\lambda = -\log(1-\rho)/T$. If B is placebo, we can compute an estimate of λ_B based on knowledge of the cumulative incidence of the disease on no therapy, and then compute the hazard rate λ_A on the experimental treatment assuming a specific reduction in risk. For instance, if λ_B is estimated to be 0.05, then if we wish to detect a 33 percent reduction in risk, $\lambda_A = (0.67)(0.05) = 0.0335$.

Under uniform recruitment over $[0, R]$ and variable follow-up over $(T - R, T]$ with no losses to follow-up, Lachin (1981) shows that

$$\begin{aligned} \phi(\lambda) &= \lambda^2 / \Pr(\text{event}|R, T, \lambda) \\ &= \lambda^2 \left(1 - \frac{e^{-\lambda(T-R)} - e^{-\lambda T}}{\lambda R} \right)^{-1}. \end{aligned} \tag{2.9}$$

In this and subsequent expressions, $\Pr(\text{event}|R, T, \lambda)$ is the probability of the event in a cohort recruited over R years and followed over T years with hazard rate λ. Thus the power of the test and the required sample size depends on these probabilities. When substituted into (2.8), this yields the required sample size needed to provide a given number of events in each group. These required numbers of events are virtually identical for other study plans specified by the values of R and T.

Lachin and Foulkes (1986) derive the expression with an adjustment for losses to follow-up under the assumption that losses to follow-up are random in each group and time to loss to follow-up is independent of the survival or event times. They consider the special case where losses are exponentially distributed with hazard rates η_A and η_B for groups A and B, respectively. Assuming uniform recruitment, equation (2.8) becomes

$$n = \frac{(\text{term1} + \text{term2})^2}{\mu_1^2 Q(1 - Q)},$$

where

$$\begin{aligned} \text{term1} &= z_{\alpha/2} \left((1 - Q)\phi(\bar{\lambda}, \eta_A) + Q\phi(\bar{\lambda}, \eta_B) \right)^{1/2}, \\ \text{term2} &= z_\beta \left((1 - Q)\phi(\lambda_A, \eta_A) + Q\phi(\lambda_B, \eta_B) \right)^{1/2}, \end{aligned}$$

and

$$\phi(\lambda, \eta) = \lambda^2 \left(\frac{\lambda}{\eta + \lambda} \left(1 - \frac{e^{-(T-R)(\eta+\lambda)} - e^{-T(\eta+\lambda)}}{R(\eta + \lambda)} \right) \right)^{-1}. \tag{2.10}$$

Note that, when $\eta = 0$, (2.10) reduces to (2.9).

Now suppose that patient entry times are distributed as truncated exponential, as in (2.1). Lachin and Foulkes (1986) show that this entry distribution yields

$$\phi(\lambda) = \lambda^2 \left(1 + \frac{\gamma e^{-\lambda T}(1 - e^{(\lambda - \gamma)R})}{(1 - e^{-\gamma R})(\lambda - \gamma)} \right)^{-1}. \tag{2.11}$$

If we employ the exponential entry distribution in conjunction with exponentially distributed losses to follow-up, we obtain

$$\phi(\lambda, \eta, \gamma) = \lambda^2 \left(\frac{\lambda}{\eta + \lambda} + \frac{\lambda \gamma e^{-(\lambda+\eta)T}(1 - e^{(\lambda+\eta-\gamma)R})}{(1 - e^{-\gamma R})(\lambda + \eta)(\lambda + \eta - \gamma)} \right)^{-1}. \tag{2.12}$$

Equation (2.12) reduces to (2.11) if $\eta = 0$.

Of course, survival may not follow an exponential distribution; other failure distributions include the Weibull and lognormal, for instance. Similar formulas could have been derived under other failure time distributions. However, it is unlikely that the investigators will have some knowledge of the form of the survival distribution *a priori*, and hence these computations can only be considered an approximation based on our current knowledge.

2.6.4 Adjustment for noncompliance

As we discussed in Section 2.3.2, compliance is not an issue if one is interested in the clinical effectiveness of a therapy, *i.e.*, the effectiveness in the general population that includes subjects who may not comply with their prescribed regimen. If a study is of pharmacologic efficacy, such noncompliant subjects are often terminated from further follow-up, or their data excluded from analysis. Of course this admits the strong potential for bias. However, if the objective of the study is true effectiveness, then under an intent-to-treat design, all such noncompliant subjects would continue to be followed and their outcome data used in all analyses. In this case, noncompliance can severely compromise power relative to an intent-to-treat effectiveness trial in which all patients are fully compliant. Similar considerations apply when inappropriate patients are entered into the trial, such as those who may be misdiagnosed, who would not benefit from treatment even if the patient were fully compliant.

To see this, let $Q = 0.5$ and let the proportion of noncompliant or inappropriate patients in the experimental group be ω and let the hazard rate be λ_A. These noncompliant patients are then assumed to have the same hazard rate as the control group. Also, assume that noncompliant patients in the control group (*e.g.*, placebo) will continue to have the same hazard rate (*e.g.*, there is no placebo effect), given by λ_B. Then μ_1^* is the expected treatment effect under noncompliance, given by $\mu_1^* = (1 - \omega)\lambda_A + \omega\lambda_B - \lambda_B = (1 - \omega)\mu_1$. Assuming that the asymptotic variances are similar in the two groups, we obtain the following adjustment from equation (2.8):

$$n^* = n/(1 - \omega)^2.$$

Note that this is quite a substantial adjustment. A noncompliance rate of 10 percent will require randomizing 23 percent more patients, Note again that this assumes that all n subjects are followed and scheduled.

In most clinical trials, the hazards specified, and the corresponding risk reduction, are specified in terms of the overall rates in the general population, recognizing that some fraction will be noncompliant. In that case, noncompliance is already allowed for in the estimates. In other cases, however, it may be desirable to specify the hazards and the risk reduction assuming 100 percent compliance. Then an adjustment such as the above could be used to factor for noncompliance, assuming complete follow-up. However, this is a severe adjustment because it assumes that a noncompliant subject has the same hazard as a control subject, regardless of how long the subject may have actually complied before becoming noncompliant, and assuming that biologically any exposure to the experimental treatment short of complete 100 percent compliance has no effect. These are implausible assumptions and thus the above is a worst case adjustment.

2.6.5   Additional considerations

While sample size calculations are a required element of proposals and protocols for randomized clinical trials, it is important to note the deficiencies of this approach. First, the formulas derived depend on unknown parameters whose values must be guessed. While the objective is to describe the sample size required to provide a desired level of power to detect a difference between groups that is clinically relevant, the actual computation requires specification of other unknown parameters. For example, in the case of two normal means, we must rely on a specification of the variance; in the case of a survival trial, we must specify the incidence of death or progression in the control group. In the latter case, for a placebo-controlled clinical trial, a substantial *placebo effect* might be present, and so these guesses are likely to be wrong. For example, suppose the hazard rate in the placebo is expected to be 0.05, but is really 0.04 due to a positive placebo effect, and we wished to detect $\mu_1 = 0.02$. With the placebo effect, we would really need to detect $\mu_1 = 0.01$, which would require a four-fold increase in sample size to detect with the same power. If our guesses are too far off, we could severely overestimate power and wind up with negative results for an effective experimental therapy. It is important to emphasize that such guesses should be conservative and calculations should be conducted over a range of values. Although it might be tempting to be as economical as possible in determining the number of subjects for an expensive clinical trial, this approach is foolhardy.

Second, these computations rely on a population model whereby individuals are assumed to be sampled at random from respective populations. Later, we introduce another approach to conducting a test of significance that is based on a randomization model that considers the probabilities of treatment assignment and their covariances, if any. Randomization models have advantages over population models, and if a randomization model is to be adopted for the final analysis of a trial, then sample size calculations based on a population may not be correct and should be viewed only as an approximation. But one can consider every aspect of sample size computation an approximation, because one must guess the underlying distribution and the underlying variability. In later chapters we will discuss the distinction between population and randomization models for the randomized clinical trial and how this might affect power.

2.7   PROBLEMS

2.1 Write a protocol for a hypothetical clinical trial. The trial will consist of a new therapy for a known disease versus a placebo. Search the medical literature for information on similar studies on the disease. Such studies should provide information on estimated incidence rates, loss to follow-up rates, information on primary outcome measures, and follow-up schedule. The protocol should include eligibility criteria, primary and secondary outcomes, study time and considerations of fixed versus variable duration, statistical analysis philosophy, and numbers of subjects randomized.

2.2 Derive equations (2.2) and (2.3) from equation (2.1).

2.3 Prove equation (2.6) from first principles. Give an intuitive explanation as to why the z_β term is unaffected for a two-sided test.

2.4 Consider independent and identically distributed observations from a Poisson distribution with rate parameter λ. The maximum likelihood estimator of λ is $\hat{\lambda} = T/n$, where T is the total number of events in n units, such as T epileptic seizures in n patient-years of exposure. Now consider two groups with parameters λ_A and λ_B with sample sizes Qn and $(1 - Q)n$, respectively, $0 < Q < 1$.
a. Derive the basic expression relating sample size and power for the test of difference between the two groups, using the formula in (2.6).
b. Consider a study to compare a drug versus placebo in the treatment of epileptics. What parameter will have to be estimated from prior knowledge?

2.5 Consider the case of two simple proportions with expectations π_A and π_B. We wish to plan a study to assess healing with an investigational drug (A) and placebo (B).
a. Derive the basic expression relating sample size and power for the two-sided test of difference of probabilities of healing between the two groups. Assume $n_A = Qn$ and $n_B = (1 - Q)n$.
b. Prior studies suggest that the control healing rate is on the order of 20 percent. Investigators believe that a minimal, clinically-meaningful increase in healing on the experimental therapy is 5 percent. For 80 percent power, compute the number of patients needed, assuming equal allocation ($Q = 0.5$).
c. From part (b), investigate the changes in n that occur with changing the allocation proportions.

2.6 For a clinical trial comparing two normal means, as presented in Section 2.6.2, suppose the standard deviation on treatment A is σ_A and the standard deviation on treatment B is σ_B, where $\sigma_A \neq \sigma_B$. Show that, for fixed n, the value of Q that maximizes power is given by $Q^* = \sigma_A/(\sigma_A + \sigma_B)$. (Allocating according to the ratio of the standard deviations is called *Neyman allocation*. This result implies that equal allocation does not maximize power when the two treatments have different standard deviations.)

2.7 The benign prostatic hyperplasia trial is a variable follow-up trial with $R = 2$ and $T = 6$, designed with four treatment groups: placebo (group I), finasteride (group II), doxazosin (group III), and combination finasteride and doxazosin (group IV). The primary outcome is three comparisons: I versus II, I versus III, and I versus IV with respect to a time-to-progression outcome. Compute the number of randomized subjects needed for these comparisons for a 50 percent reduction in risk, when the incidence rate over 5 years is assumed to be 25 percent. Make the following assumptions: $\alpha = 0.05$ (two-sided), 80 percent power, exponential incidence, and uniform recruitment. Build in a 10 percent exponential loss to follow-up rate over the five years.

2.8 REFERENCES

BLACKWELL, D. AND HODGES, J. L. (1957). Design for the control of selection bias. *Annals of Mathematical Statistics* **28** 449–460.

COCHRAN, W. G. AND COX, G. M. (1957). *Experimental Designs*. Wiley, New York.

DIABETES PREVENTION PROGRAM RESEARCH GROUP. (2002). Reduction in the incidence of type 2 diabetes with lifestyle intervention or metformin. *New England Journal of Medicine* **346** 393–403.

DONNER, A. (1984). Approaches to sample size estimation in the design of clinical trials – a review. *Statistics in Medicine* **3** 199–214.

ECHT, D. S., LIEBSON, P. R., MITCHELL, L. B., PETERS, R. W., OBIAS-MANNO, D., BARKER, A. H. FOR THE CARDIAC ARRHYTHMIA SUPPRESSION TRIAL. (1991). Mortality and morbidity in patients receiving encainide, flecainide, or placebo. *New England Journal of Medicine* **324** 781–788.

FLEMING, T. R. AND DEMETS, D. L. (1996). Surrogate end points in clinical trials: are we being misled? *Annals of Internal Medicine* **125** 605–613.

INTERNATIONAL CHRONIC GRANULOMATOUS DISEASE COOPERATIVE STUDY GROUP. (1991). A controlled trial of interferon gamma to prevent infection in chronic granulomatous disease. *New England Journal of Medicine* **324** 509–516.

JONES, B. AND KENWARD, M. G. (1989). *Design and Analysis of Crossover Trials*. Chapman and Hall, London.

LACHIN, J. M. (1981). Introduction to sample size determination and power analysis for clinical trials. *Controlled Clinical Trials* **2** 93–113.

LACHIN, J. M. (2000). Statistical considerations in the intent-to-treat principle. *Controlled Clinical Trials*, **21** 167–189.

LACHIN, J. M. AND FOULKES, M. A. (1986). Evaluation of sample size and power for analyses of survival with allowance for nonuniform patient entry, losses to follow-up, noncompliance, and stratification. *Biometrics* **42** 507–519.

LAKATOS, E. (1988). Sample sizes based on the log-rank statistic in complex clinical trials. *Biometrics* **44** 229–241.

LEWIS, E. J., HUNSICKER, L. G., BAIN, R. P., ROHDE, R. D., AND THE COLLABORATIVE STUDY GROUP (1993). The Effect of Angiotensin-Converting-Enzyme Inhibition on Diabetic Nephropathy. *New England Journal of Medicine* **329** 1456–1462.

PRENTICE, R. L. (1989). Surrogate endpoints in clinical trials: definition and operational criteria. *Statistics in Medicine* **8** 431–440.

SAMUEL-CAHN, E. AND WAX, Y. (1986). A sequential test for comparing two infection rates in a randomized clinical trial, and incorporation of data accumulated after stopping. *Biometrics* **42** 99–108.

3

Randomization for Balancing Treatment Assignments

3.1 INTRODUCTION

Thus far we have talked quite loosely about randomization as the toss of a fair coin, but simple coin-tossing is rarely employed in a clinical trial. One can distinguish four classes of randomization procedures: *complete randomization, restricted randomization, covariate-adaptive randomization,* and *response-adaptive randomization.*

Let $T_1, ..., T_n$ be a sequence of random treatment assignments, where $T_i = 1$ if the patient is assigned to treatment A and $T_i = 0$ if the patient is assigned to treatment B, $i = 1, ..., n$. *Complete randomization* is simple coin-tossing, in which case, $T_1, ..., T_n$ are independent and identically distributed Bernoulli random variables with $p = \Pr(T_i = 1) = 1/2, i = 1, ..., n$. In *restricted randomization procedures,* $T_1, ..., T_n$ are dependent, with variance-covariance matrix given by $\Sigma_T \neq (1/4)I$. Restricted randomization is employed when it is desired to have equal numbers of patients assigned to each treatment group (*i.e.,* balancing the treatment assignments), and this will be the topic of this chapter. *Covariate-adaptive randomization* is used when it is desired to ensure balance between treatment arms with respect to certain known covariates. Treatment assignments will depend on the covariate values of patients. Finally, *response-adaptive randomization* is used when ethical considerations make it undesirable to have equal numbers of patients assigned to each treatment arm. In response-adaptive randomization, the treatment assignments depend upon previous patient responses to treatment. Covariate-adaptive and response-adaptive randomization will be treated in later chapters. The four types of randomization procedures are progressively more complicated, from a statistical point of view, due to the increased complexity of the dependence structure.

In this chapter, we discuss complete randomization and restricted randomization procedures for balancing treatment assignments. Friedman, Furberg, and DeMets (1981, p. 41) present two main arguments for equal allocation to treatment groups. The first is that power is maximized when allocation is equal. The second is that equal allocation is consistent with the concept of equipoise that should exist at the beginning of the trial. Many clinical trialists disagree, in principle, with these arguments, and we will explore alternative arguments in later in chapters on response-adaptive randomization. However, it should be noted that most clinical trials today do employ restricted randomization procedures to achieve balance, and these arguments have become rooted, to some extent, in the culture of clinical trials.

We now present the principal randomization tools to achieve balance among the treatment groups. A thorough probabilistic analysis of these randomization procedures is required. In particular, the conditional expectation of assignment, given all previous assignments, will define the procedure. The unconditional variance-covariance structure of the treatment assignments is used to develop the theoretical susceptibility to bias of each procedure (Chapters 5 and 6) as well as to determine the distribution of randomization-based inferential tests (Chapter 7).

3.2   THE BALANCING PROPERTIES OF COMPLETE RANDOMIZATION

When treatment assignments are independent Bernoulli random variables with success probability $1/2$, we have complete randomization. Complete randomization has some very nice properties, in that certain types of bias are minimized. For example, there can be no selection bias with complete randomization, since it is equally likely to guess the next treatment assignment correctly or incorrectly. However, there is a disadvantage to complete randomization that makes it unattractive in practice: there is a non-negligible probability of some imbalances between treatments and a small probability of severe imbalances. In fact, the theory of probabilities of large deviations should serve as a warning when using Bernoulli sequences for randomization in small to moderate samples.

Let $N_A(i) = \sum_{j=1}^{i} T_j, i = 1, ..., n$ so that $N_A(i)$ is the number of patient randomized to treatment A after i patients have been randomized. Let $N_B(i) = i - N_A(i)$. Then by the central limit theorem for a binomial random variable, $N_A(n)$ is asymptotically normal with mean $n/2$ and asymptotic variance $n/4$. Letting $D_n = N_A(n) - N_B(n) = 2N_A(n) - n$, we see that D_n is asymptotically normal with mean 0 and variance n. We can use $|D_n|$ as one measure to describe the degree of imbalance between treatment groups. For $r > 0$,

$$\Pr(|D_n| > r) \cong 2\{1 - \Phi(r/\sqrt{n})\}, \tag{3.1}$$

where Φ is the standard normal distribution function. One can use this formula to determine the degree to which complete randomization is subject to imbalances of size r, for a large sample trial with n patients (see Problem 3.2). Table 3.1 gives percentiles of the distribution of $|D_n|$ for various values of n. For example, when

Table 3.1   *Percentiles of the distribution of* $|D_n|$ *for complete randomization.*

n	0.33	0.25	0.10	0.05	0.025
50	6.9	8.1	11.6	13.9	15.9
100	9.7	11.5	16.5	19.6	22.4
200	13.8	16.3	23.3	27.7	31.7
400	19.5	23.0	32.9	39.2	44.8
800	27.6	32.5	46.5	55.4	63.4

$n = 50$, there is a five percent chance of an imbalance of ± 13.9 or worse. This corresponds to an excess of 6.95 beyond the expected 25 in either group, or an imbalance of 36.1 percent versus 63.9 percent. When $n = 400$ there is a five percent chance of an imbalance of ± 39.2, corresponding to a degree of imbalance of 45.1 percent versus 54.9 percent. To many, an imbalance of this degree would be of no concern.

While some imbalances will likely occur, apart from cosmetic concerns, the important question is whether these imbalances compromise the statistical properties of the study. Regardless of the final sample sizes, balanced or not, the resulting estimate of the treatment group difference will still be unbiased. While an imbalance will decrease the precision of the estimator, this effect will be slight for moderate imbalances.

Likewise, an imbalance will decrease the power of statistical test, but again, the effect is slight for moderate imbalances. For example, consider the example of Section 2.6.2, the comparison of two normal means with equal variances. Power can be computed using equation (2.7). We draw the power curves across values of Q for $n = 50, 100, 200$ in Figure 3.1, where $\sigma = 1$ and $|\mu_1| = 0.5$. For large n, the curve flattens at the top, indicating that there is little loss of power for Q between 0.30 and 0.70. As shown in Table 3.1, the probability of a large imbalance following complete randomization is very small. It is even smaller with restricted randomization designs.

3.3   RANDOM ALLOCATION RULE

Because there is a significant probability of an imbalance between treatments for clinical trials employing complete randomization, one may wish to impose a restriction that the final allocation be exactly equal between the two treatment groups. This can be accomplished by using a *random allocation rule* [cf. Lachin (1988)], provided that the investigator has control over the total number of subjects to be randomized. While this is feasible in animal studies or small phase II studies, it is not always possible in a phase III clinical trial.

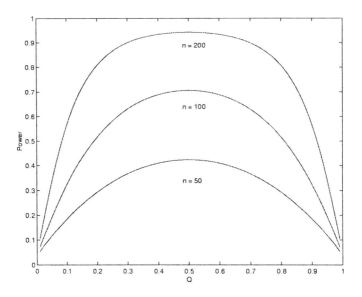

Fig. 3.1   *Power curves for the comparison of two normal means across values of Q.*

Let $\mathcal{F}_n$ be the set of treatment assignments through the first n stages of the randomization process, *i.e.*, $\mathcal{F}_n = \{T_1, ..., T_n\}$. [For a formal mathematical treatment, $\mathcal{F}_n$ is a *sigma-algebra* and $\mathcal{F}_0$ is the trivial sigma algebra. See Chapters 13–15.] Note that $\Pr(T_n = 1) = E(T_n)$. Then the random allocation rule is defined by the following allocation probabilities:

$$E(T_j | \mathcal{F}_{j-1}) = \frac{\frac{n}{2} - N_A(j-1)}{n - (j-1)}, j = 2, ..., n, \tag{3.2}$$

and $E(T_1) = 1/2$. For example, if patient 50 is ready to be randomized in a clinical trial of $n = 100$ patients, and thus far the allocation has been 28 to A and 21 to B. Then patient 50 will receive treatment A with probability $22/51$.

One can immediately see some problems with this rule. First, once $n/2$ patients have been assigned to one treatment, all further treatment allocations are deterministic, and hence absolutely predictable, and selection bias can result. Second, at some stage in the middle of the trial, there could be significant treatment imbalances. If patients entering the trial are heterogeneous with respect to some important covariate related to outcome (*e.g.*, if there is a time trend), then imbalances between treatment groups with respect to that covariate may result. For example, suppose $n_1 < n$ patients have been randomized in the clinical trial. Conditional on n_1, $N_A(n_1)$ follows a hypergeometric distribution with mean $n_1/2$ and variance $n_1(n - n_1)/4(n - 1)$. The exact probability of imbalance of $N_A(n_1)$ to $N_B(n_1)$ can be obtained from Fisher's exact test for the resulting 2×2 table with cells $N_A(n_1), N_B(n_1), n/2 - N_A(n_1)$, and $n/2 - N_B(n_1)$. Asymptotically, the distribution of $N_A(n_1)$ can be approximated by a normal distribution. Let $D_{n_1,n} = N_A(n_1) - N_B(n_1)$. For $r > 0$,

$$\Pr(|D_{n_1,n}| > r) \cong 2\left\{1 - \Phi\left(r\sqrt{(n-1)/n_1(n-n_1)}\right)\right\}. \tag{3.3}$$

Clearly, this function is maximized when the $n_1/n = 0.5$, or only half the planned allocations are completed.

One can think of the random allocation rule in terms of an urn model. Suppose an urn contains $n/2$ balls of type A and $n/2$ balls of type B. Each time a patient is ready to be randomized, a ball is drawn and *not* replaced, and the corresponding treatment is assigned. This continues until the urn is depleted. One can easily see that this leads to the allocation rule in (3.2).

With the urn formulation, we can see that random allocation rule produces $\binom{n}{n/2}$ equally likely permutations of $n/2$ A's and $n/2$ B's. Therefore, the unconditional probability of treatment assignment, given by $E(T_j)$, can be found by thinking of the jth element of the $\binom{n}{n/2}$ permutation sequences. Of those sequences, half of the jth elements are A and half are B. Since each sequence is equally likely,

$$E(T_j) = \left(\binom{n}{n/2}\bigg/2\right)\bigg/\binom{n}{n/2} = 1/2.$$

Since $T_j^2 = T_j$, it follows that $\mathrm{Var}(T_j) = E(T_j) - \{E(T_j)\}^2 = 1/4$, as for complete randomization. The differences between complete randomization and restricted randomization are completely specified by $\mathrm{cov}(T_i, T_j)$, which is 0 for complete randomization. For the random allocation rule, we can compute, for $i < j$,

$$\begin{aligned}
E(T_iT_j) &= \Pr(T_i = 1, T_j = 1) \\
&= \Pr(T_j = 1 | T_i = 1) \cdot E(T_i) \\
&= \frac{n/2 - 1}{n - 1} \cdot \frac{1}{2} \\
&= \frac{n - 2}{4(n - 1)}.
\end{aligned}$$

Consequently,

$$\mathrm{cov}(T_i, T_j) = \frac{n - 2}{4(n - 1)} - \frac{1}{4} = \frac{-1}{4(n - 1)}. \tag{3.4}$$

3.4 TRUNCATED BINOMIAL DESIGN

An alternate way of assigning exactly $n/2$ patients to each treatment is to randomly allocate each according to the toss of a coin until one treatment has been assigned

$n/2$ times; all subsequent patients will receive the opposite treatment. Blackwell and Hodges (1957) refer to this as the *truncated binomial design*. We can describe the design by the rule

$$
\begin{aligned}
E(T_j|\mathcal{F}_{j-1}) &= 1/2, && \text{if} & \max\{N_A(j-1), N_B(j-1)\} &< n/2, \\
&= 0, && \text{if} & N_A(j-1) &= n/2, \\
&= 1, && \text{if} & N_B(j-1) &= n/2,
\end{aligned}
$$

for n even. As in the random allocation rule, the tail of the randomization sequence will be entirely predictable and there can be serious imbalances during the course of the trial.

We can quantify the number of subjects in the tail whose treatment assignment is deterministic using discrete distribution theory (Blackwell and Hodges, 1957). Let X be the random number of subjects in the tail. Then $X = x$ can occur if the $n/2$th A assignment or the $n/2$th B assignment occurs for patient $n - x$, for n even. These events have probability

$$
\frac{1}{2^{n-x}} \left(\begin{array}{c} n - x - 1 \\ \frac{n}{2} - 1 \end{array} \right),
$$

according to the negative binomial distribution. Therefore,

$$
\Pr(X = x) = \frac{1}{2^{n-x-1}} \left(\begin{array}{c} n - x - 1 \\ \frac{n}{2} - 1 \end{array} \right), x = 1, ..., n/2, \tag{3.5}
$$

a truncated negative binomial distribution. From this distribution, one can derive

$$
E(X) = \frac{n}{2^n} \left(\begin{array}{c} n \\ n/2 \end{array} \right) \tag{3.6}
$$

and

$$
E(X^2) = n - E(X) \tag{3.7}
$$

(Problem 3.3). The results (3.6) and (3.7) will be needed later in Chapter 6.

We now compute the unconditional mean, variance, and covariance of the treatment assignment indicators. This can be accomplished by conditioning on the random variable $\tau = \min\{i : \max(N_A(i), N_B(i)) = n/2\}$. It is clear that $\Pr(N_A(\tau) = n/2) = \Pr(N_B(\tau) = n/2) = 1/2$. Conditional on $N_A(\tau) = n/2$, we compute the probability of assignment to A as follows:

$$
\begin{aligned}
E(T_j) &= EE(T_j|\tau) \\
&= E \Pr(T_j = 1|\tau) \\
&= 1/2 \cdot \Pr(\tau \ge j) + 0 \cdot \Pr(\tau < j). \tag{3.8}
\end{aligned}
$$

Similarly, conditional on $N_B(\tau) = n/2$, we compute

$$
\begin{aligned}
E(T_j) &= E \Pr(T_j = 1|\tau) \\
&= 1/2 \Pr(\tau \ge j) + 1 \cdot \Pr(\tau < j). \tag{3.9}
\end{aligned}
$$

Since (3.8) and (3.9) each have probability $1/2$, we compute the unconditional expectation as

$$
\begin{aligned}
E(T_j) &= 1/4 \cdot \Pr(\tau \geq j) + 1/2 \cdot \Pr(\tau < j) + 1/4 \cdot \Pr(\tau \geq j) \\
&= 1/2.
\end{aligned}
$$

Therefore, $\mathrm{Var}(T_j) = 1/4$.

It remains to compute $\mathrm{cov}(T_i, T_j)$. Again, we break the problem into two pieces by conditioning first on $N_A(\tau) = n/2$. For $i < j$,

$$
\begin{aligned}
E(T_i T_j) &= EE(T_i T_j | \tau) \\
&= E\Pr(T_i = 1, T_j = 1 | \tau) \\
&= 1/4 \cdot \Pr(\tau \geq j) + 0 \cdot \Pr(\tau < j).
\end{aligned}
$$

Similarly, conditioning on $N_B(\tau) = n/2$, we obtain

$$
\begin{aligned}
E(T_i T_j) &= E\Pr(T_i = 1, T_j = 1 | \tau) \\
&= 1/4 \cdot \Pr(\tau \geq j) + 1/2 \cdot \Pr(i \leq \tau < j) + 1 \cdot \Pr(\tau < i).
\end{aligned}
$$

Unconditioning, we see that

$$
\begin{aligned}
E(T_i T_j) &= 1/4 \cdot \Pr(\tau \geq j) + 1/4 \cdot \Pr(i \leq \tau < j) + 1/2 \cdot \Pr(\tau < i) \\
&= 1/4 + 1/4 \cdot \Pr(\tau < i).
\end{aligned}
$$

Consequently,

$$
\mathrm{cov}(T_i, T_j) = 1/4 \cdot \Pr(\tau < i). \tag{3.10}
$$

The distribution of τ comes from (3.5), as $\Pr(X = x) = \Pr(\tau = n - x)$, and hence

$$
\mathrm{cov}(T_i, T_j) = \frac{1}{4} \sum_{k=n-i+1}^{n/2} \frac{1}{2^{n-k-1}} \binom{n-k-1}{\frac{n}{2}-1}. \tag{3.11}
$$

Note that $\mathrm{cov}(T_i, T_j) = 0$ if $i < (n+2)/2$.

While the truncated binomial design and the random allocation rule both yield $\binom{n}{n/2}$ permutations of As and Bs, the sequences will not be equiprobable for the truncated binomial design. This is demonstrated in Table 3.2 for $n = 4$.

3.5 PERMUTED BLOCK DESIGNS

Complete randomization, the random allocation rule, and the truncated binomial design can all result in severe imbalances at some point during the trial. This is particularly undesirable if there is a time-heterogeneous covariate that is related to treatment outcome, because imbalances in treatment assignments can then lead to

Table 3.2 Six permutation sequences for $n = 4$ with probabilities under truncated binomial randomization (each sequence has probability $1/6$ under random allocation).

Sequence	Probability
$AABB$	$1/4$
$ABAB$	$1/8$
$ABBA$	$1/8$
$BABA$	$1/8$
$BAAB$	$1/8$
$BBAA$	$1/4$

imbalances in those important covariates. To avoid this, *permuted block designs* are often used to ensure balance throughout the course of the clinical trial, by imposing a balance restriction at various stages in the trial. Permuted block designs are reviewed by Zelen (1974).

For the permuted block design, we establish M blocks containing $m = n/M$ patients, where M and n/M are positive integers, and within block i, $m/2$ patients are assigned to treatment A and $m/2$ patients are assigned to treatment B. To ensure balance, a random allocation rule is typically used within each block (although one could also use a truncated binomial design), where the total number of patients is m instead of n. When permuted blocks are used, at M stages during the course of the trial, we achieve balanced allocation. The maximum imbalance at any time during the trial is given by $\max_j |D_j| = m/2$.

One could also use a truncated binomial design to achieve complete balance within each block. In this case the underlying assignment sequence probabilities, and the covariance matrix of the assignments, would be different from those of a random allocation rule.

In the extreme case, $M = n/2$ and every pair randomized is balanced. However, this procedure requires every even randomization to be deterministic, and hence selection bias is easy to occur, unless the pairs are first identified and then randomized as a set. For these reasons, block sizes larger than 2 are generally employed, and investigators should be masked to the block size selected.

Some biostatisticians advocate a *variable block design* in which block sizes are selected randomly. In this case each of the M blocks has m_i patients, $i = 1, ..., M$, where the size of the ith block is selected at random. The motivation for selecting random block sizes is to reduce the chance of selection bias. However, in Chapter 6 we will see that under a commonly used model for selection bias, use of random block sizes yields virtually no reduction in such bias.

3.6   EFRON'S BIASED COIN DESIGN

Efron (1971) developed the *biased coin design* in order to balance treatment assignments. Let D_n be an increasing function of $N_A(n)$ such that $D_n = 0$ if $N_A(n) = n/2$ [for example, $D_n = N_A(n) - N_B(n) = 2N_A(n) - n$]. He suggests allocating with the following rule. Define a constant $p \in (0.5, 1]$.

$$
\begin{aligned}
E(T_j|\mathcal{F}_{j-1}) &= 1/2, &&\text{if} \quad D_{j-1} = 0, \\
&= p, &&\text{if} \quad D_{j-1} < 0, \\
&= 1 - p, &&\text{if} \quad D_{j-1} > 0.
\end{aligned}
$$

We can use a symmetry argument to determine the unconditional probability of assignment to A, $E(T_j)$. By symmetry (and since $D_0 = 0$), we have

$$
\Pr(D_{j-1} < 0) = \Pr(D_{j-1} > 0),
$$

and both equal $1/2$ if $j - 1$ is odd. Hence, for $j - 1$ odd, we have

$$
E(T_j) = p\Pr(D_{j-1} < 0) + (1 - p)\Pr(D_{j-1} > 0) = 1/2.
$$

For $j - 1$ even, we have

$$
\Pr(D_{j-1} < 0) + \Pr(D_{j-1} > 0) = 2\Pr(D_{j-1} < 0) = 1 - \Pr(D_{j-1} = 0),
$$

and hence

$$
\begin{aligned}
E(T_j) &= \Pr(D_{j-1} = 0)/2 + p\Pr(D_{j-1} < 0) + (1 - p)\Pr(D_{j-1} > 0) \\
&= \Pr(D_{j-1} = 0)/2 + [1 - \Pr(D_{j-1} = 0)]/2 \\
&= 1/2.
\end{aligned}
$$

Using the theory of random walks, we can measure the degree of imbalance, $|D_n| = |2N_A(n) - n|$. We have the following transition probabilities:

$$
\Pr\left(|D_{n+1}| = 1 \,\middle|\, |D_n| = 0\right) = 1,
$$

and, for a positive integer j,

$$
\Pr\left(|D_{n+1}| = j - 1 \,\middle|\, |D_n| = j\right) = p,
$$

$$
\Pr\left(|D_{n+1}| = j + 1 \,\middle|\, |D_n| = j\right) = 1 - p.
$$

This yields the following random walk matrix for $|D_n|$:

$$
P = \begin{bmatrix}
0 & 1 & 0 & 0 & 0 & \cdots \\
p & 0 & 1-p & 0 & 0 & \cdots \\
0 & p & 0 & 1-p & 0 & \cdots \\
0 & 0 & p & 0 & 1-p & \cdots \\
\vdots & \vdots & \vdots & \vdots & \vdots &
\end{bmatrix}.
$$

We can solve the steady-state equations, by solving for the left eigenvector of P [cf. Karlin and Taylor (1975, p. 86)]. These equations are given by

$$
\begin{aligned}
\pi_0 &= p\pi_1, \\
\pi_1 &= \pi_0 + p\pi_2, \\
\pi_2 &= (1-p)\pi_1 + p\pi_3, \\
&\vdots \quad \vdots
\end{aligned}
$$

and $\pi_0 + \pi_1 + \pi_2 + \cdots = 1$. The solution is

$$
\begin{aligned}
\pi_0 &= \frac{r-1}{2r}, \\
\pi_j &= \frac{(r+1)(r-1)}{2r^{j+1}}, j \geq 1,
\end{aligned}
\tag{3.12}
$$

where $r = p/(1-p)$ (Problem 3.4). Since $|D_n|$ can take only odd or even values as n is odd or even, the Markov chain has period 2, and the π's must be doubled [cf. Ross (1983, p. 111)]. We can obtain the limiting balancing property as

$$
\begin{aligned}
\lim_{m \to \infty} \Pr(|D_{2m}| = 0) &= 2\pi_0 = 1 - \frac{1}{r}, \\
\lim_{m \to \infty} \Pr(|D_{2m+1}| = 1) &= 2\pi_1 = 1 - \frac{1}{r^2}.
\end{aligned}
$$

(Note that for odd n, the minimum imbalance is 1.) Obviously, as $p \to 1$, we achieve perfect balance, but such a procedure is deterministic. When $p = 2/3$, we have probability of $1/2$ of achieving perfect balance for even n and probability $3/4$ with odd n for large n.

Soares and Wu (1982) modified Efron's procedure by considering a level of imbalance that would be unacceptable, and then imposing a deterministic treatment assignment (by setting $p = 0$) to counter the imbalance. Their design, which they named the *big stick rule*, is given by

$$
\begin{aligned}
E(T_j|\mathcal{F}_{j-1}) &= 1/2, \quad \text{if} \quad |D_{j-1}| < c, \\
&= 0, \quad \text{if} \quad D_{j-1} = c, \\
&= 1, \quad \text{if} \quad D_{j-1} = -c.
\end{aligned}
$$

The degree of imbalance is given by the constant c, which is fixed in advance. A similar procedure, originally developed by Larry Shaw, was used in the National Cooperative Gallstone Study in which a proportionate degree of imbalance, $D_{j-1}/(j-1)$, was employed in lieu of an absolute difference to define the acceptable degree of imbalance in the above expression (Lachin, Marks, Schoenfield, et al., 1981).

Chen (1999) introduced a hybrid of the big stick rule and Efron's biased coin design, calling it the *biased coin design with imbalance intolerance*. The rule is given by

$$
\begin{aligned}
E(T_j|\mathcal{F}_{j-1}) &= 1/2, & \text{if} \quad D_{j-1} = 0, \\
&= 0, & \text{if} \quad D_{j-1} = c, \\
&= 1, & \text{if} \quad D_{j-1} = -c, \\
&= p, & \text{if} \quad 0 < D_{j-1} < c, \\
&= 1-p, & \text{if} \quad -c < D_{j-1} < 0.
\end{aligned}
$$

The design induces a random walk on the state space $\{0, ..., c\}$ with reflecting barriers 0 and c. Asymptotic balancing properties for this stochastic process are derived in Chen (1999).

3.7 WEI'S URN DESIGN

Using Efron's biased coin design, the bias of the coin, p, is constant, regardless of the degree of imbalance. Wei (1977, 1978) developed an *adaptive biased coin design*, where the probabilities of assignment adapt according to the degree of imbalance. One convenient model that adapts these probabilities is an urn model. A review of urn randomization is found in Wei and Lachin (1988).

For the urn design, initially an urn contains α balls of each of two types, A and B. When a patient is ready to be randomized, a ball is drawn and *replaced*. If the ball is type A, treatment A is assigned to the patient and β type B balls are added to the urn. If the ball is type B, treatment B is assigned to the patient and β type A balls are added to the urn. In this way, the urn composition is skewed to increase the probability of assignment to the treatment that has been selected least often thus far. As with other designs, the sequence of assignments can be conducted in advance of patient enrollment. The urn design is denoted $UD(\alpha, \beta)$, and has the allocation rule

$$
\begin{aligned}
E(T_j|\mathcal{F}_{j-1}) &= \frac{\alpha + \beta N_B(j-1)}{2\alpha + \beta(j-1)}, \, j \geq 2; \\
E(T_1|\mathcal{F}_0) &= 1/2.
\end{aligned}
\tag{3.13}
$$

If $\alpha = 0$, the first treatment assignment occurs with probability $1/2$. Note that the $UD(\alpha, 0)$ design is complete randomization. For the $UD(0, 1)$, we have the following simple allocation rule:

$$
E(T_j|\mathcal{F}_{j-1}) = \frac{N_B(j-1)}{j-1}, j \geq 2.
\tag{3.14}
$$

We can show that the unconditional probability of assignment to A is $1/2$ using induction, using (3.14). First note that $E(T_1) = E\{N_A(1)\} = 1/2$. Assume $E\{N_A(j-1)\} = (j-1)/2$ Then

$$
\begin{aligned}
E\{N_A(j)\} &= EE\{T_j|\mathcal{F}_{j-1}\} + E\{N_A(j-1)\} \\
&= 1 - \frac{E\{N_A(j-1)\}}{j-1} + \frac{j-1}{2} \\
&= \frac{j}{2}.
\end{aligned}
$$

Thus $E(T_j) = 1/2$.

We can examine the transition probabilities for the degree of imbalance, $|D_n|$, of the $UD(\alpha, \beta)$ as follows. Without loss of generality, assume $N_A(n) - N_B(n) = j$, for positive integer j. Then, since $N_A(n) + N_B(n) = n$, we have $N_A(n) = (j+n)/2$. So

$$
\begin{aligned}
\Pr\left(|D_{n+1}| = j - 1 \Big| |D_n| = j\right) &= \frac{\text{number of type B balls in urn}}{\text{total number of balls in urn}} \\
&= \frac{\alpha + \beta N_A(n)}{2\alpha + \beta n} \\
&= \frac{1}{2} + \frac{j\beta}{2(2\alpha + n\beta)}, \\
\Pr\left(|D_{n+1}| = j + 1 \Big| |D_n| = j\right) &= \frac{1}{2} - \frac{j\beta}{2(2\alpha + n)\beta}, \\
\Pr\left(|D_{n+1}| = 1 \Big| |D_n| = 0\right) &= 1.
\end{aligned}
$$

So, asymptotically, $UD(\alpha, \beta)$ tends to complete randomization. Wei (1977) uses the recursive formula

$$
\begin{aligned}
&\Pr(|D_{n+1}| = j) \\
&= \Pr\left(|D_{n+1}| = j \Big| |D_n| = j - 1\right) \Pr(|D_n| = j - 1) \\
&\quad + \Pr\left(|D_{n+1}| = j \Big| |D_n| = j + 1\right) \Pr(|D_n| = j + 1) \quad (3.15)
\end{aligned}
$$

to find the unconditional distribution of $|D_n|$. Wei (1977, p. 384) tabulates these values for $j \leq 10$.

Wei (1978b) shows that $N_A(n)$ is asymptotically normal with mean $n/2$ and asymptotic variance $(\alpha + \beta)n/4(3\beta - \alpha)$, provided $3\beta > \alpha$. Consequently, D_n is asymptotically normal with mean 0 and asymptotic variance $(\alpha + \beta)n/(3\beta - \alpha)$. We can then compute, for integer r,

$$
\Pr(|D_n| > r) \cong 2\left\{1 - \Phi\left(r\sqrt{\frac{(3\beta - \alpha)}{(\alpha + \beta)n}}\right)\right\}. \quad (3.16)
$$

This can be compared directly with equation (3.1) for complete randomization. In Table 3.3, we evaluate percentiles of the asymptotic imbalance distribution for the urn design. We see that, for the $UD(0, 1)$, the urn design has a lower probability of imbalance than complete randomization asymptotically (the entries in Table 3.1 are simply divided by $\sqrt{3}$). The same is true for any value of β when $\alpha = 0$. Note that when $\alpha = \beta$, the asymptotic probability of imbalance is the same as for complete randomization.

Other urn designs have been proposed in the literature for use in randomized clinical trials. Several will be discussed in the context of response-adaptive randomization

Table 3.3 *Percentiles of the distribution of* $|D_n|$ *for Wei's urn design.*

(α, β)	n	0.33	0.25	0.10	0.05	0.025
(0,1)	50	4.0	4.7	6.7	8.0	9.2
(0,1)	100	5.6	6.6	9.5	11.3	12.9
(0,1)	200	8.0	9.4	13.4	16.0	18.3
(0,1)	400	11.2	13.3	19.0	22.6	25.9
(0,1)	800	15.9	18.8	26.9	32.0	36.6
(1,3)	50	4.9	5.8	8.2	9.8	11.2
(1,3)	100	6.9	8.1	11.6	13.9	15.8
(1,3)	200	9.7	11.5	16.4	19.6	22.4
(1,3)	400	13.8	16.3	23.3	27.7	31.7
(1,3)	800	19.5	23.0	32.9	39.2	44.8

in later chapters. Another restricted randomization design for achieving balance is the *Ehrenfest urn model* proposed by Chen (2000). In this urn, c (even) balls are arranged in two urns, labelled A and B, with $c/2$ balls in each urn. The c balls are equally likely to be drawn. One draws one of the c balls at random. If the ball came from urn A, treatment A is assigned and the ball is replaced in urn B. If it came from urn B, treatment B is assigned and it is replaced in urn A. Balancing is achieved because it is more likely to draw a ball from the urn which contains more balls, thus reducing the composition of that urn by 1. Unlike Wei's urn, the Ehrenfest urn maintains a constant number of balls.

The Ehrenfest urn was originally proposed in physics to obtain equilibrium between two isolated bodies. The urn induces a Markov chain on the state space $\{0, ..., c\}$ with reflecting barriers at 0 and c, and, as such, is directly comparable to Efron's biased coin design with imbalance control described in the previous section. Chen (2001) does a comparison in terms of asymptotic balancing properties, and finds that the Ehrenfest urn is more effective than the biased coin design with imbalance control (where c is the same value in both) when, for the biased coin design, $1/c < p < 1/2$. When $0 < p < 1/c$, the biased coin design has better balancing properties.

3.8 GENERALIZED BIASED COIN DESIGNS

Wei's urn design and Efron's biased coin design can be thought of as special cases of a more general framework which we will call *generalized biased coin designs* [see Smith (1984)]. Define a function $\phi(j) = \phi(N_A(j), N_B(j))$ such that

$$E(T_j|\mathcal{F}_{j-1}) = \phi(j - 1).$$

The function $\phi(j)$ describes a large class of designs including Efron's, which takes the form

$$
\begin{aligned}
\phi(j) &= 1/2, && \text{if } N_A(j) = N_B(j), \\
&= p, && \text{if } N_A(j) < N_B(j), \\
&= 1-p, && \text{if } N_A(j) > N_B(j),
\end{aligned}
$$

and Wei's, which can be written in the form

$$
\phi(j) = \frac{1}{2} - \frac{\beta\{N_A(j) - N_B(j)\}}{4\alpha + 2\beta j}. \tag{3.17}
$$

Wei (1978a) generalized (3.17), proposing

$$
\phi(j) = p\left(\frac{N_A(j) - N_B(j)}{j}\right), \tag{3.18}
$$

where p is a nonincreasing function satisfying $p(x) + p(-x) = 1$. Smith (1984) proposed a class of designs depending on a positive parameter ρ, given by

$$
\phi(j) = \frac{N_B(j)^\rho}{N_A(j)^\rho + N_B(j)^\rho}. \tag{3.19}
$$

This rule corresponds to

$$
p(x) = \frac{(1-x)^\rho}{(1+x)^\rho + (1-x)^\rho}
$$

in equation (3.18). If $\rho = 1$, we have Wei's urn design with $\alpha = 0$. If $\rho = 0$, we have complete randomization ($\phi(j) = 1/2$ for all j). Smith favors the design with $\rho = 5$.

3.9   COMPARISON OF BALANCING PROPERTIES

Table 3.4 gives a simulation comparison of four designs used for balance: complete randomization, Efron's biased coin design ($p = 2/3$), Wei's $UD(0,1)$, and Smith's design with $\rho = 5$. One can see that complete randomization does not balance as well as the three restricted randomization procedures, and Efron's and Smith's designs are very close in terms of the bias and variability. Wei's is slightly more variable.

3.10 $K > 2$ TREATMENTS

For clinical trials of more than two treatments, most of these randomization procedures generalize quite readily. Complete randomization becomes a simple multinomial probability generator with K equally likely outcomes. The random allocation rule can be thought of as an urn with n/K balls representing each treatment. Truncated binomial randomization becomes a multi-stage process whereby K-treatment

Table 3.4   *Simulated mean and variance of $N_A(n)/n$ for four different randomization sequences, $n = 50$, based on 10,000 replications.*

Randomization procedure	$E(N_A(n)/n)$	Var$(N_A(n)/n)$
Complete	0.5004	0.0049
$BCD(2/3)$	0.4998	0.0004
$UD(0,1)$	0.5001	0.0016
Smith ($\rho = 5$)	0.4999	0.0005

complete randomization is used, and each treatment is subsequently dropped when the n/Kth patient is assigned to that treatment, until only one treatment is left. All subsequent patients are then assigned to that treatment.

Efron's biased coin design does not generalize so easily. Since allocation is based on the value of D_n, some generalized measure of imbalance among the K treatments would have to be developed. Wei's urn design admits an easier generalization. For the $UD(\alpha, \beta)$ design, the urn contains α balls representing each treatment initially. Then β balls are added for each other treatment after an assignment is made. For the $UD(0, 1)$, the probability that the jth assignment is to treatment i, given the previous $j - 1$ assignments is given by

$$\frac{j - 1 - N_i(j - 1)}{(j - 1)(K - 1)}. \tag{3.20}$$

This reduces to (3.14) when $K = 2$.

Wei, Smythe, and Smith (1986) describe a wide class of designs for K treatments that are an extension of (3.18). Suppose the desired allocation proportions for the K groups is $\boldsymbol{\xi} = (\xi_1, ..., \xi_K)$ (with all $\xi_i = 1/K$ equal unless one wishes unbalanced allocation; see Section 3.11). Let $\boldsymbol{p} = (p_1(\boldsymbol{x}), ..., p_{K-1}(\boldsymbol{x}))$ be continuous functions of some $(K - 1) \times 1$ vector $\boldsymbol{x}$, where p_i is the probability that patient j will be assigned to treatment i and $p_K = 1 - \sum_{k=1}^{K-1} p_k$. Let $\boldsymbol{N}(j) = (N_1(j), ..., N_{K-1}(j))$ be the number of patients assigned to treatment $i = 1, ..., K - 1$ after j patients have been assigned. Then typically, $p_1, ..., p_K$ will be a function of $\boldsymbol{x} = \boldsymbol{N}(j-1)/(j-1)$. The $p_1, ..., p_K$ are assumed to satisfy the following relationship:

$$\text{if } x_i \geq \xi_i, \text{ then } p_i(\boldsymbol{x}) \leq \xi_i, j = 1, ..., K. \tag{3.21}$$

For some special cases of this general procedure, consider first complete randomization. Here $p_i(\boldsymbol{x}) = 1/K$ for all i, so that future assignments are independent of previous assignments. For the generalization of Wei's urn in (3.20), we have $p_i(\boldsymbol{x}) = (1 - x_i)/(K - 1)$. Note that Efron's biased coin design is not a special case of (3.21) because $\boldsymbol{p}$ is not continuous for each i.

We have the following important limiting result from Wei, Smythe, and Smith (1986):

$$p_i \left(\frac{N(j-1)}{j-1} \right) \to \xi_i, \tag{3.22}$$

in probability, as $j \to \infty$, for $i = 1, ..., K$. So this general rule for K treatments tends asymptotically to the desired allocation.

3.11   RESTRICTED RANDOMIZATION FOR UNBALANCED ALLOCATION

Some statisticians [e.g., Peto (1978)] have advocated that, under certain conditions, clinical trials should randomize with fixed unequal allocation probabilities. For example, one might allocate in a 2:1 ratio of intervention to control. Sometimes favoring the experimental therapy is warranted in trials of potentially great public health benefit, such as when testing a new AIDS therapy, where patients may be reluctant to have only a 50 percent chance of receiving the new therapy. Such unequal allocation procedures can improve recruitment. One must remember that the experimental therapy may also be harmful, and hence unequal allocation could subject more patients to a harmful therapy. There are also cases where widespread knowledge about the control therapy exists and more understanding is needed about the experimental therapy. Although the study may lose some sensitivity, there may be gain in terms of information about the toxicity and patient responses to the experimental therapy. This argument was used to justify 2:1 allocation in an oncology trial (Cocconi, Bella, Zironi, et al., 1994). Such decisions could be controversial and should be made in the context of careful power assessments.

In some cases, an optimal allocation ratio different from $1 : 1$ will maximize power (see, for example, Problem 2.6). This is discussed further in Chapter 10. Another instance where an unbalanced design is statistically desirable is when the principal analysis is a set of multiple comparisons of $K - 1$ treatments versus a single control using the Dunnett (1955) procedure. In this case, the *square-root rule* provides an optimal set of allocation ratios $(K - 1)^{1/2} : 1 : 1 : \cdots : 1$. For three treatments, the allocation proportions are $(0.414, 0.293, 0.293)$. Rarely, however, is the subset of $K - 1$ comparisons employed as the principal analysis in lieu of an overall test of equality of treatment means.

Generally, an unbalanced design, if employed, is justified on the grounds of ethics or cost. In later chapters, we describe the use of response-adaptive randomization, which dynamically alters the allocation probabilities to reflect the accruing data on the trial, putting more patients on the treatment performing better thus far. While different from fixed unbalanced allocation, where the allocation probabilities are determined in advance of the trial, the rationale for response-adaptive randomization is much the same.

In practice, restricted randomization can be altered to produce a fixed unbalanced allocation. For complete randomization and the truncated binomial design, a biased coin can be tossed with the desired allocation proportion for treatment A. For the random allocation rule, let the urn initially contain the desired proportion of balls of each treatment. While Efron's biased coin design does not readily admit a simple generalization, we can modify Wei's urn design to allow for unequal allocation. Let the desired allocation for A be $0 < Q < 1$. Then the urn initially contains $Q\alpha$ balls representing treatment A and $(1 - Q)\alpha$ balls representing treatment B. If a type A ball is drawn, $(1 - Q)\beta$ balls are added of type B. If a type B ball is drawn, $Q\beta$ balls of type A are added.

3.12 PROBLEMS

3.1 Repeat Figure 3.1 for the standard Z-test of a difference of two proportions.

3.2 Analyze the balancing property of complete randomization and Wei's urn design ($\alpha = 0, \beta = 1$) theoretically using the normal approximations given in equations (3.1) and (3.16). For imbalances $\Pr(|D_n|/n > r/n)$, and $r/n = 0.05, 0.10, 0.20$, graph the probability of imbalance versus sample size (for $n = 1$ to 100) with graphics software. Superimpose the three curves (for the three values of r/n) for each procedure. Interpret and compare results with Table 3.4.

3.3 Derive equations (3.6) and (3.7).

3.4 Derive the solution to the steady-state equations for Efron's biased coin design, given in equations (3.12).

3.13 REFERENCES

BLACKWELL, D. AND HODGES, J. L. (1957). Design for the control of selection bias. *Annals of Mathematical Statistics* **28** 449–460.

CHEN, Y-P. (1999). Biased coin design with imbalance intolerance. *Communications and Statistics – Stochastic Models* **15** 953–975.

CHEN, Y-P. (2000). Which design is better? Ehrenfest urn versus biased coin. *Advanced in Applied Probability* **32** 738–749.

COCCONI, G., BELLA, M., ZIRONI, S., ALGERI, R., DI COSTANZO, F., DE LISI, V., LUPPI, G., MAZZOCCHI, B., RODINO, C., AND SOLDANI, M. (1994). Flourouracil, doxorubicin, and mitomycin combination versus PELF chemotherapy in advanced gastric cancer: a prospective randomized trial of the Italian Oncology Group for Clinical Research. *Journal of Clinical Oncology* **12** 2687–2693.

DUNNETT, C. W. (1955). A multiple comparison procedure for comparing several treatments with a control. *Journal of the American Statistical Association* **50** 1096–1121.

EFRON, B. (1971). Forcing a sequential experiment to be balanced. *Biometrika* **58** 403–417.

FRIEDMAN, L. M., FURBERG, C. D., AND DEMETS, D. L. (1981). *Fundamentals of Clinical Trials*. Wright PSG, Boston.

KARLIN, S. AND TAYLOR, H. M. (1975). *A First Course in Stochastic Processes*. Academic Press, Boston.

LACHIN, J. M. (1988). Properties of simple randomization in clinical trials. *Controlled Clinical Trials* **9** 312–326.

LACHIN, J. M., MARKS, J., SCHOENFIELD L. J., THE PROTOCOL COMMITTEE AND THE NCGS GROUP. (1981). Design and methodological considerations in the National Cooperative Gallstone Study: a multi-center clinical trial. *Controlled Clinical Trials* **2** 177–230.

PETO, R. (1978). Clinical trial methodology. *Biomedicine* **28** 24–36.

ROSS, S. M. (1983). *Stochastic Processes*. Wiley, New York.

SMITH, R. L. (1984). Sequential treatment allocation using biased coin designs. *Journal of the Royal Statistical Society B* **46** 519–543.

SOARES, J. F. AND WU, C. F. J. (1982). Some restricted randomization rules in sequential designs. *Communications in Statistics – Theory and Methods* **12** 2017–2034.

WEI, L. J. (1977). A class of designs for sequential clinical trials. *Journal of the American Statistical Association* **72** 382–386.

WEI, L. J. (1978a). The adaptive biased coin design for sequential experiments. *Annals of Statistics* **6** 92–100.

WEI, L. J. (1978b). An application of an urn model to the design of sequential controlled clinical trials. *Journal of the American Statistical Association* **73** 559–563.

WEI, L. J. AND LACHIN, J. M. (1988). Properties of the urn randomization in clinical trials. *Controlled Clinical Trials* **9** 345–364.

WEI, L. J., SMYTHE, R. T., AND SMITH, R. L. (1986). K-treatment comparisons with restricted randomization rules in clinical trials. *Annals of Statistics* **14** 265–274.

ZELEN, M. (1974). The randomization and stratification of patients to clinical trials. *Journal of Chronic Diseases* **27** 365–375.

4

Balancing on Known Covariates

4.1 INTRODUCTION

Chapter 1 pointed out the likelihood of confounders in biomedical studies. In any clinical trial, there are *covariates* or *prognostic factors* of interest besides the treatment effect. Some covariates are known in advance to be important risk factors that are associated with the outcome of a patient. For instance, in trials of heart disease, relevant covariates may be cholesterol, blood pressure, age, or gender. In a randomized study, the objective is to equalize the distribution of such factors within each treatment group so as to minimize biases due to covariate imbalances. The most common covariate causing such concern in multi-center clinical trials is the clinical center, since clinics usually differ with respect to the demographic and clinical characteristics of their patient populations, and adherence to the protocol and various procedures. Thus an imbalance in the numbers randomized to each group within a clinic, such as 60% to A in clinic 1 and 30% to A in clinic 2, may bias the results of the study. The goal is to balance allocation of treatments within each clinic. By so doing, the distribution of the covariate (clinic in this case) is equalized within each treatment group. Other covariates often known in advance to be important include gender, age, race, and medical baseline measurements important in the assessment of the disease.

Although randomization tends to mitigate the possibility of serious covariate imbalances among the treatment groups, it is not unusual for imbalances to occur in practice. In this event, stratified analyses or regression modelling can be used to adjust for important covariates in a post hoc analysis. Another alternative is to implement a design that ensures balance on specified covariates in the trial. In this chapter we discuss randomization techniques that tend to balance treatments within the discrete levels, or *strata*, of known covariates. We describe the relative merits of stratified randomization. We then describe other approaches that are designed to systematically control or prevent covariate imbalances.

4.2 STRATIFIED RANDOMIZATION

In a randomized study, *stratification* has interchangeably been used to refer to either a stratified randomization (often termed *pre-stratification*), or a stratified-adjusted analysis (often termed *post-stratification*) with or without a stratified randomization. To avoid ambiguity, the distinction is drawn between a stratified randomization and a stratified-adjusted analysis, each as described below.

In a stratified randomization (pre-stratification), subjects are grouped according to covariate values prior to randomization, and subjects are then randomized within strata. Within each stratum, a separate randomization sequence is employed. For example, consider a study stratified by clinic (say 5 in number) and gender, with 10 total strata defined jointly by the covariates clinic with 5 categories and gender with 2. In this case, a separate randomization sequence would be employed for each gender and for each clinic, 10 sequences in all. A female subject recruited by clinic 3 would be randomized using the "clinic 3 and female" randomization sequence. If a restricted randomization sequence is employed, such as an urn design, then the probability of assignment of this subject would depend only on the number of prior assignments to A and B among females recruited by clinic 3, and not on the prior treatment assignments to males recruited by clinic 3, nor on the prior assignments to males or females recruited by other clinics. In this case, the randomization sequences within the 10 strata would be accomplished by using 10 separate urns.

Stratification is also used to refer to post-randomization stratification in the analysis whereby the subjects are grouped within strata or subgroups according to one or more patient characteristics. In the above example, the analysis might be performed using the 10 strata defined on the joint basis of the covariates "clinic" and "gender". The first stage of the analysis is to compare treatments A versus B separately within each strata. This is also called a *subgroup analysis*. In the second stage, various methods might then be used to perform a combined test, by pooling the results of all the strata or subgroups in some way, so as to provide an aggregate test over strata or subgroups.

For the above example of a randomization stratified by clinic and gender, consider the simplest case of two treatment groups A versus B and a binary response (*e.g.* "healed" versus "not healed"). Within each clinic-gender stratum, a 2×2 table can be formed to assess the treatment-response association within that stratum, expressed as an odds ratio. To then assess the aggregate treatment-response association over all 10 strata combined, the Mantel-Haenszel procedure could be applied. This provides a stratified-adjusted estimate of the common odds ratio, obtained as a linear combination of the within-strata odds ratios; and an aggregate stratified-adjusted test of association. The analysis effectively adjusts for the stratum effects since treatments

A and *B* are compared within strata and then the results are combined over strata [*cf.* Lachin (2000)].

An alternate strategy is to simply combine patients and responses over all strata into a single 2 × 2 table for which *A* and *B* are then compared in a single analysis, ignoring any strata, even those used as a basis for a stratified randomization. In this case, the stratified randomization is effectively ignored in the analysis. This *pooled analysis* is also called a *combined, unstratified,* or *unadjusted analysis.*

Note that a stratified-adjusted analysis can be performed for any groupings of subjects regardless of whether the randomization was stratified according to those groupings. Conversely, an unstratified analysis can be performed, even though the randomization may have been stratified by other factors. Thus an initial consideration might be the relative efficiency of a stratified-adjusted analysis with a stratified versus non-stratified randomization, and the relative efficiency of a stratified versus unstratified analysis of a study that employed stratified randomization. These will be explored in detail in Chapter 8.

The gains from stratification were early recognized in sample surveys where it was shown that a stratified analysis improves the precision of estimators. However, it is principally the stratified analysis which eliminates bias, for which a stratified randomization is not necessary. Thus, it should not be surprising that a stratified randomization tends to improve the efficiency of estimators and power of tests in a small trial, say for $n < 100$, but has negligible advantage in large trials. This issue has been discussed by various authors using various models.

Stratification is often considered to be an essential feature of randomization (*cf.* Zelen (1974)), but there has been significant controversy as to the relative statistical merits of stratified randomization versus a stratified analysis following unstratified randomization. With unstratified randomization, the probability of covariate imbalances decreases as the sample size increases and is usually of little consequence. Also, for moderate to large sample sizes, unstratified randomization affords slightly less statistical power than does stratified randomization, but the difference is negligible. The main consideration is that a stratified analysis, not a stratified randomization, adjusts for any bias due to a covariate imbalance. Thus, it is often recommended that the randomization for a clinical trial should be stratified only on those factors considered absolutely necessary to ensure the integrity of the study [*cf.* Friedman, Furburg and DeMets (1985), Peto, Pike, Armitage, *et al.* (1976)].

In many studies, differences among clinics are the major source of heterogeneity in the outcome measures. Further, since a clinic frequently may withdraw (or be dropped) from a study, it is desirable that such withdrawal should not affect the validity of the overall randomization plan. For these reasons, it is also generally advocated that randomization in a multi-center trial should be stratified by clinic.

4.3   TREATMENT IMBALANCES IN STRATIFIED TRIALS

A common misconception is that stratified randomization promotes greater balance between the numbers of treatment assignments to A and B within each stratum and thus overall. Unfortunately, this is not always so.

Since some randomization procedures, including complete randomization, Efron's biased coin design, and Wei's urn, do not force balance between treatments (except asymptotically), there is a positive probability that imbalances will occur within individual strata when stratified randomization is performed. Let N_{iA} be the number of patients assigned to treatment A in stratum $i, i = 1, ..., s$ Then $N_A(n) = \sum_{i=1}^{s} N_{iA}$. Asymptotically, $N_A(n)$ should be approximately $n/2$. However, for finite samples, with a large number of small strata, imbalances are additive across strata, and can result in an overall imbalance of some significance. This is less likely to occur when there are small numbers of large strata.

Often a permuted block design is used within each stratum to ensure balance. This is called *stratified blocked randomization*. With the random allocation rule or permuted block design, there is no imbalance within strata or in aggregate as long as all blocks are filled. However, if some blocks are not filled, a treatment imbalance can occur. Since an unstratified randomization risks at most one unfilled block, whereas a s-strata randomization risks at most s unfilled blocks, the probability of a treatment imbalance is greater in a clinical trial with stratified randomization.

In stratified blocked randomization, one must be careful to limit the stratification variables and the number of strata within each to a minimum, representing only the most important variables and levels. For instance, in a multi-center trial with 15 participating institutions, stratifying by clinic, gender, race (3 levels), and age (2 levels) leads to $15 \times 2 \times 3 \times 2 = 180$ strata. Unless the trial is extremely large, some strata will have very few patients.

Hallstrom and Davis (1988) describe the probability of aggregate imbalances in a trial when using stratified blocked randomization. Suppose n patients are assigned with equal probability to treatments A and B, within s strata. Each of the s strata are balanced by using permuted blocks, and the block size in the ith stratum is $b_i, i = 1, ..., s$. Let N_i be the number of patients assigned in the last block of stratum i, and A_i be the number assigned to treatment A, $1 \leq A_i \leq N_i \leq b_i$. Define $D_i = N_i - 2A_i$ to be the imbalance in the ith stratum. Conditional on N_i, A_i has a hypergeometric distribution, with expectation $N_i/2$ and variance $N_i(b_i - N_i)/4(b_i - 1)$. Then $E(D_i|N_i) = 0$ and $E(D_i) = 0$. We can then derive

$$\text{Var}(D_i) = \frac{E\{N_i(b_i - N_i)\}}{b_i - 1} \qquad (4.1)$$

(see Problem 4.1). Summing over independent strata, the total imbalance in the trial is given by

$$D = \sum_{i=1}^{s} D_i.$$

Using equation (4.1), we can determine the effect of block size and the number of strata on the variability of D, provided we have some information on the distribution of N_i. Hallstrom and Davis (1988) consider two models. The first model assumes that the expected number of patients in each stratum is large relative to the block size. In this case, it is reasonable to assume that N_i follows a discrete uniform distribution on the support $\{1, ..., b_i\}$. Then $E(N_i) = (b_i + 1)/2$ and $E(N_i^2) = (b_i + 1)(2b_i + 1)/6$. From (4.1), we can compute $\text{Var}(D_i) = (b_i + 1)/6$. So, under the uniform model, we have

$$\text{Var}(D) = \frac{\sum_{i=1}^{S} b_i + S}{6}. \tag{4.2}$$

The normal approximation can be used to compute

$$\Pr(|D| > d) \cong 2\left\{1 - \Phi\left(\frac{d}{\sqrt{\text{Var}(D)}}\right)\right\} \tag{4.3}$$

for various values of b_i. Such an exercise is useful in planning studies.

For example (see Hallstrom and Davis (1988, p. 378)), the Cardiac Arrhythmia Suppression Trial (CAST) was planned with a total of 4200 patients and 270 strata. Using equation (4.2) with $b_i = 4$, we find that $\text{Var}(D) = 225$. Then by equation (4.3), we can compute $\Pr(|D| > 30) = 0.05$. Such a difference is small compared to the number of patients, and would not be of concern. The maximum imbalance is $\sum_{i=1}^{K} b_i/2 = 540$.

One can perform similar analyses if Wei's urn is used instead of a permuted block design, using asymptotic formulas for the variance of an imbalance given in Section 3.7.

4.4   COVARIATE-ADAPTIVE RANDOMIZATION

In the previous sections, we have assumed that a set of s strata is defined on the basis of one or more covariates and a separate randomization is performed within each stratum. An entirely different approach would be to determine the treatment assignment of a new subject to minimize the covariate imbalances within treatment groups. This approach has been called *adaptive stratification* or *covariate-adaptive randomization*.

4.4.1   Zelen's rule

Zelen's (1974) rule uses a pre-assigned randomization sequence (which could be generated by complete randomization or some restricted randomization design) ignoring strata. Let $N_{ik}(n)$ be the number of patients in stratum $i = 1, ..., S$ on treatment $k = 1, 2$ $(1 = A, 2 = B)$. When patient $n + 1$ in stratum i is ready to be randomized, one computes $D_i(n) = N_{i1}(n) - N_{i2}(n)$. For an integer c, if $|D_i(n)| < c$, then the patient is randomized according to schedule, otherwise, the patient receives the opposite treatment. Zelen proposes $c = 2, 3$, or 4. He also proposes randomizing the value of c for each new patient.

4.4.2   The Pocock-Simon procedure

In a similar vein to Zelen's rule, Pocock and Simon (1975) proposed a *covariate-adaptive randomization* procedure. Let $N_{ijk}(n), i = 1, ..., I, j = 1, ..., n_i, k = 1, 2 (1 = A, 2 = B)$, be the number of patients in stratum j of covariate i on treatment k after n patients have been randomized. (In our previous notation, $\prod_{i=1}^{I} n_i = s$ is the total number of strata in the trial.) Suppose the $(n+1)$th patient to be randomized is a member of strata $r_1, ..., r_I$ of covariates $1, ..., I$. Again, we define D to be a difference metric; in this case, let $D_i(n) = N_{ir_i1}(n) - N_{ir_i2}(n)$. We then take a sum over weighted strata defined by $D(n) = \sum_{i=1}^{I} w_i D_i(n)$, where w_i are weights chosen depending on which covariates are deemed of greater importance. If $D(n)$ is less than $1/2$, then the weighted difference measure indicates that B has been favored thus far for that set, $r_1, ..., r_I$, of strata and the patient $n + 1$ should be assigned with higher probability to treatment A, and vice-versa, if $D(n)$ is greater than $1/2$. Pocock and Simon suggest biasing a coin with

$$p = \frac{c^* + 1}{3} \qquad (4.4)$$

and implementing the following rule: if $D(n) < 1/2$, assign the next patient to treatment A with probability p; if $D(n) > 1/2$, assign the next patient to treatment A with probability $1 - p$; and if $D(n) = 1/2$, assign the next patient to treatment A with probability $1/2$, where $c^* \in [1/2, 1]$.

Note that if $c^* = 1$, we have a rule very similar to Efron's biased coin design of Section 3.6. If $c^* = 2$, we have the deterministic *minimization method* proposed by Taves (1974) (see also Simon (1979)). Many other rules could be considered, all derivatives of Zelen's rule and Taves's minimization method with a biased coin twist to give added randomization; Efron (1980) describes one such rule and applies it to a clinical trial in ovarian cancer.

Pocock and Simon generalize their covariate-adaptive randomization procedure for more than two treatments by considering a general metric $D_i^k(n), k = 1, ..., K$, which could be the standard deviation of the $N_{ir_ik}(n)$'s, and a weighted sum $D^k(n) = \sum_{i=1}^{I} w_i D_i^k(n)$. The D^k's are then ordered from smallest to largest, and a corresponding set of probabilities $p_1 \geq p_2 \geq \cdots \geq p_K$ is determined such that $\sum_{k=1}^{K} p_k = 1$. The value $p_k, k = 1, ..., K$, is then the probability that patient $n + 1$ with strata $r_1, ..., r_I$ will be assigned to treatment k. Pocock and Simon suggest the following functional form:

$$p_k = c^* - \frac{2(Kc^* - 1)}{K(K + 1)} k, k = 1, ..., K. \qquad (4.5)$$

Note that (4.5) reduces to (4.4) for $K = 2$.

4.4.3 Wei's marginal urn design

Wei (1978) described the use of an urn model for covariate-adaptive randomization. When the number of covariates is such that the resulting number of strata is large and the stratum sizes are small, using a separate urn in each stratum can result in imbalances in treatment assignments within strata. Let n_i be the number of categories for the ith of I stratification factors considered jointly, such that there are $s = \prod_{i=1}^{I} n_i$ unique strata. Instead of using s urns, one for each unique stratum, Wei proposed using an urn for each category of each covariate, for a total of $\sum_{i=1}^{I} n_i$ urns. For a given new subject with covariate values $r(1), ..., r(I)$, the treatment group imbalances within each of the corresponding urns is examined. The one with the greatest imbalance is used to generate the treatment assignment. A ball from that urn is chosen and then replaced. Then β balls representing the opposite treatment are added to the urns corresponding to that patient's covariate values. Wei called this a *marginal urn* because it tends to balance treatment assignments within each category of each covariate marginally, and thus also jointly.

4.5 OPTIMAL DESIGN BASED ON A LINEAR MODEL

The rules of Zelen and Pocock and Simon in the preceding sections are arbitrary in the sense that they are developed intuitively rather than based on some optimal criterion. While one can simulate these procedures for different parameter values and find appropriate designs to fit certain criteria, none of the designs has been shown to be optimal. Instead of concerns about balance of treatment assignments across strata, one can take an entirely different approach and find an allocation rule that minimizes the variance of the estimated treatment effect in the presence of covariates. Such a rule would necessarily require the specification of a model linking the covariates and the treatment effect. Begg and Iglewicz (1980) and Atkinson (1982) select a standard linear regression model. Here we follow Atkinson's development. We begin with the classical regression model, given by

$$E(Y_i) = x_i'\beta, i = 1, ..., n,$$

where the Y_i's are independent responses with $\mathrm{Var}(Y) = \sigma^2 I$ and x_i includes a treatment indicator and selected covariates of interest. Then $\mathrm{Var}(\hat{\beta}) = \sigma^2 (X'X)^{-1}$, where $X'X$ is the $p \times p$ dispersion matrix from n observations.

For the construction of optimal designs, we wish to find the n points of experimentation at which some function is optimized (in our case we will be finding the optimal sequence of n treatment assignments). The dispersion matrix evaluated at these n points is given by $M(\xi_n) = X'X/n$, where ξ_n is the n-point design. It is convenient, instead of thinking of n points, to formulate the problem in terms of a measure ξ (which in this case a frequency distribution) over a design region Ξ.

Since an important goal of clinical trials is to estimate a treatment effect, possibly adjusting for important covariates, Atkinson formulates the optimal design problem as a design that minimizes, in some sense, the variance of $A'\beta$, where A is an $s \times p$

matrix of contrasts, $s < p$. One possible criterion is Sibson's (1974) D_A-optimality that maximizes

$$|A'M^{-1}(\xi)A|^{-1}. \tag{4.6}$$

Other criteria could also be applied. Atkinson compares the D_A criterion to standard D-optimality, which maximizes the log determinant of M. Ball, Smith, and Verdinelli (1993) investigate the Bayesian D-optimality criterion, where a Bayesian prior distribution is assumed for β, and the procedure maximizes the expectation (with respect to the prior distribution) of the log determinant of M.

For any multivariable optimization problem, we compute the directional derivative of the criterion. In the case of the D_A criterion in (4.6), we can derive the directional (Frechet) derivative as

$$d_A(x, \xi) = x'M^{-1}(\xi)A(A'M^{-1}(\xi)A)^{-1}A'M^{-1}(\xi)x.$$

By the classical equivalence theorem of Kiefer and Wolfowitz (1960), the optimal design ξ^* that maximizes the criterion (4.6) then satisfies the following equations:

$$\sup_{x \in \Xi} d_A(x, \xi) \le s, \ \forall \xi \in \Xi \tag{4.7}$$

and

$$\sup_{x \in \Xi} d_A(x, \xi^*) = s \tag{4.8}$$

[Kiefer and Wolfowitz (1960); see Atkinson and Donev (1992) for further details.]

In the model with covariates, we have

$$E(Y) = x_1\beta_1 + x_2'\beta_2,$$

where x_1 is the treatment indicator vector and x_2 is a vector of important covariates. In this case, if we are interested in estimating the treatment effect in the presence of covariates, $A' = [A_1' : 0]$ with A_1 identifying the treatment differences. This formulation can be simplified with two treatments, but the optimal design that satisfies (4.7) and (4.8) must be determined numerically.

Such a design is optimal for estimating linear contrasts of β, but the solution will provide only an allocation ratio on each of K treatments, without incorporating the sequential nature of a clinical trial. Assume n patients have already been allocated, and the resulting n-point design is given by ξ_n. Atkinson proposes a sequential design which allocates the $(n + 1)$th patient to the treatment $k = 1, ..., K$ for which $d_A(k, \xi_n)$ is a maximum. However this design is deterministic.

In order to randomize the allocation, Atkinson suggests biasing a coin with probabilities

$$p_k = \frac{d_A(k, \xi_n)}{\sum_{k=1}^{K} d_A(k, \xi_n)} \tag{4.9}$$

and allocating to treatment k with the corresponding probability. With two treatments, $k = 1, 2$ $(1 = A, 2 = B)$, we have $s = 1$, $A' = [-1, 1]$, and the probability of assigning treatment A is given by

$$p = \frac{d_A(1, \xi_n)}{d_A(1, \xi_n) + d_A(2, \xi_n)}. \tag{4.10}$$

With no covariates in the model, the model becomes $E(Y) = \beta_k, k = 1, 2$, and the equations in (4.7) and (4.8) can be solved analytically. We can write (4.10) as

$$p = \frac{\{N_B(n)\}^2}{\{N_A(n)\}^2 + \{N_B(n)\}^2}, \tag{4.11}$$

where $N_A(n)$ and $N_B(n)$ are the numbers of patients assigned to treatments A and B, respectively, through n patients (Problem 4.4). Note that this is the design in equation (3.19) with $\rho = 2$. In a similar vein, Ball, Smith, and Verdinelli (1993) and Atkinson (1998a) investigate Bayesian optimality criteria.

Atkinson (1998b, 1999) compares his rule in (4.9) to Efron's biased coin, complete randomization, and stratified blocked randomization in terms of the variance of the estimated treatment effect by simulating four independent normal covariates and four correlated normal covariates. He concludes that the less randomized the design, the smaller the variance, but the greater the potential biases.

While Atkinson's approach has the advantage of incorporating a formal optimality criterion in the problem of randomizing in the presence of important covariates, it has several disadvantages. First, he relies on a linear model deals with continuous homogeneous responses, whereas many clinical trials deal with binary or surival endpoints. However, transformations can be a useful tool in establishing approximately normal responses. If not, his results are likely to be applicable for generalized linear models as well, although this has not been addressed formally, as far as we know. Second, the algorithm is computationally intensive when dealing with several covariates. Finally, Atkinson's approach is concerned with minimizing the variance of treatment contrasts in the presence of important covariates. This is not the same goal as balancing over covariates to mitigate biases. An interesting dialog on the relative importance of these goals can be found in the discussion following Atkinson (1999).

4.6 CONCLUSIONS

Stratified blocked randomization has become quite popular in today's randomized clinical trials. However, one should be aware that such designs can, in some cases, result in treatment imbalances in the trial, due to incomplete blocks in some strata. Stratification can also quickly become overwhelming if there are many important covariates in the trial. Consequently, covariate-adaptive randomization procedures have been proposed, which are variants on Efron's biased coin design. Adaptive stratification is used to ensure balance without requiring separate randomization within prespecified strata. However, such designs are largely *ad hoc* and have numerous parameters that must be specified. Simulation is a good tool to investigate the relationship of these parameters to potential biases (see Problem 4.3). Alternatively, one could employ Atkinson's optimal design approach if there is time to implement numerical optimization procedures. However, the optimal design approach relies on a specific linear model which may not be appropriate for some trials, and the goal of the procedure is to minimize the variance of the treatment effect estimator in the presence of covariates, rather than to balance over known covariates.

4.7   PROBLEMS

4.1 Derive equation (4.1).

4.2 Refer to the CAST example in Section 4.3. Suppose each clinic contains four hospitals, so that the number of strata becomes $270 \times 4 = 1080$. Determine the probability than an imbalance greater than 30 will result and the maximum possible imbalance assuming that N_i is uniform. Comment on the appropriateness of stratified blocks in this setting (Hallstrom and Davis, 1988).

4.3 Suppose you are planning a clinical trial with $n = 2400$ patients to be randomized in 12 clinical centers. Assume the probability that a sequentially entered patient is randomized in a particular clinical center is $1/12$ for each center. Simulate the balancing property $D = 2N_A(n) - n$ for the following randomization procedures:
(*i*) permuted blocks with $b_i = 10$ in each stratum;
(*ii*) Zelen's rule with $c = 2$;
(*iii*) The Pocock-Simon procedure with $c^* = 0.50, 0.75, 1.00$.
a. Which method is better in terms of $E(D)$ and $\text{Var}(D)$?
b. How does the value of c affect the covariate-adaptive procedure?
c. Compute $\text{Var}(D)$ using equation (4.2). How does it compare to the simulated value?
d. Suppose three clinics are poor recruiters, and the probability that a patient is recruited in clinics 1, 2, or 3 is $1/30$ in each and $1/10$ in each of clinics 4, ..., 12. How do the results change?

4.4 Derive equation (4.11).

4.8   REFERENCES

ATKINSON, A. C. (1982). Optimum biased coin designs for sequential clinical trials with prognostic factors. *Biometrika* **69** 61–67.

ATKINSON, A. C. (1998a). Bayesian and other biased-coin designs for sequential clinical trials. *Tatra Mountains Mathematical Publications* **17** 133–139.

ATKINSON, A. C. (1998b). Optimum experimental designs for chemical kinetics and clinical trials. In *New Developments and Applications in Experimental*

Design (FLOURNOY, N., ROSENBERGER, W. F., AND WONG, W. K., EDS.). Institute of Mathematical Statistics, Hayward, pp. 36–49.

ATKINSON, A. C. (1999). Optimum biased-coin designs for sequential treatment allocation with covariate information. *Statistics in Medicine* **18** 1741–1752.

ATKINSON, A. C. AND DONEV, A. N. (1992). *Optimum Experimental Designs.* Clarendon Press, Oxford.

BALL, F. G., SMITH, A. F. M., AND VERDINELLI, I. (1993). Biased coin designs with Bayesian bias. *Journal of Statistical Planning and Inference* **34** 403–421.

BEGG, C. B. AND IGLEWICZ, B. (1980). A treatment allocation procedure for sequential clinical trials. *Biometrics* **36** 81–90.

EFRON, B. (1980). Randomizing and balancing a complicated sequential experiment. In *Biostatistics Casebook* (MILLER, R. G., EFRON, B., BROWN, B. W., AND MOSES, L. E., EDS.). Wiley, New York, pp. 19–30.

FRIEDMAN, L. M., FURBERG, C. D., AND DEMETS, D. L. (1998). *Fundamentals of Clinical Trials.* Springer, New York.

HALLSTROM, A. AND DAVIS, K. (1988). Imbalance in treatment assignments in stratified blocked randomization. *Controlled Clinical Trials* **9** 375–382.

KIEFER, J. AND WOLFOWITZ, J. (1960). The equivalence of two extremum problems. *Canadian Journal of Mathematics* **12** 363–366.

LACHIN, J. M. (2000). *Biostatistical Methods: The Assessment of Relative Risks.* Wiley, New York.

PETO, R., PIKE, M. C., ARMITAGE, P., BRESLOW, N. E., COX, D. R., HOWARD, S. V., MANTEL, N., MCPHERSON, K., PETO, J., AND SMITH, P. G. (1976). Design and analysis of randomized clinical trials requiring prolonged observation of each patient – I: Introduction and design. *British Journal of Cancer* **34** 585–612.

POCOCK, S. J. AND SIMON, R. (1975). Sequential treatment assignment with balancing for prognostic factors in the controlled clinical trial. *Biometrics* **31** 103–115.

SIBSON, R. (1974). *D*-optimality and duality. In *Progress in Statistics* (GANI, J., SARKADI, K. AND VINCZE, J., EDS.). North-Holland, Amsterdam.

SIMON, R. (1979). Restricted randomization designs in clinical trials. *Biometrics* **35** 503–512.

TAVES, D. R. (1974). Minimization: a new method of assigning patients to treatment and control groups. *Clinical Pharmacology and Therapeutics* **15** 443–453.

WEI, L. J. (1978). An application of an urn model to the design of sequential controlled clinical trials. *Journal of the American Statistical Association* **73** 559–563.

ZELEN, M. (1974). The randomization and stratification of patients to clinical trials. *Journal of Chronic Diseases* **28** 365–375.

5

The Effects of Unobserved Covariates

5.1 INTRODUCTION

In Chapter 4, we stressed the importance of advance planning in mitigating the effects of certain known covariates that influence the primary outcome. However, the human physiology is so complex that it is simply impossible to identify every covariate that may be related to treatment outcome. One can certainly adjust for any covariates collected during the clinical trial in a post-hoc analysis, but it is likely that some important covariates may not have been collected. This is one of the great benefits of randomization: randomization tends to balance treatment assignments on these unknown covariates.

Some statisticians have argued that randomization is unnecessary. However, a simple example, adapted from Berry (1989), demonstrates its benefits. Suppose one wishes to compare the time it takes to drive to work on two different routes. Does it make a difference in our conclusions if we drive five times consecutively on one route and then five times consecutively on the other route, or if we flip a coin for 10 days to randomly to select the route? One can quickly respond that, yes, it does matter, if it snowed the first five days and there was good weather on the latter five days. Intuitively, it would seem that randomizing the route would make it more likely to have similar numbers of days of bad weather for both routes. This is the concept that randomization tends to balance treatment assignments on unknown covariates that may be related to treatment. This intuition is correct, but only asymptotically. For small samples, there is still a significant probability that an imbalance will result. For the extreme case, randomizing the experiment still leads to a probability that one route is taken only in bad weather of 1/252. Consequently, believers in randomization will talk only about "mitigating" covariate imbalances. In fact, everyone who has participated in a randomized clinical trial has witnessed covariate imbalances among treatment groups, even with carefully conducted randomization and relatively large numbers of randomized subjects. It can be presumed that such phenomena provide non-believers with some degree of satisfaction.

In this chapter, we explore some of the theoretical properties of randomization in mitigating biases from unknown covariates. We begin with simple probability statements on covariate imbalances and then examine Efron's more sophisticated model for accidental bias.

5.2 A BOUND ON THE PROBABILITY OF A COVARIATE IMBALANCE

The simplest method to analyze the probability of a covariate imbalance is to use Chebyshev's inequality. Suppose $T_1, ..., T_n$ are randomly assigned treatment indicators (again $T_i = 1$ if treatment A and 0 if treatment B) and let $\mathcal{F}_n = \{T_1, ..., T_n\}$. Assume $n_A = \sum_{i=1}^n T_i$ is fixed in advance, and $n_A = Qn, Q \in (0, 1)$. Suppose Z is some covariate of interest, and $Z_1, ..., Z_n$ are independent and identically distributed with mean μ and variance σ^2. Then $\bar{Z}_A = \sum_{i=1}^n T_i Z_i / n_A$ and $\bar{Z}_B = \sum_{i=1}^n (1 - T_i) Z_i / (n - n_A)$. A covariate imbalance between treatment groups would be represented by a tangible difference between $\bar{Z}_A$ and $\bar{Z}_B$. Using a conditioning argument, we can then derive the following:

$$
\begin{aligned}
E\{\bar{Z}_A - \bar{Z}_B\} &= EE\left\{\frac{n\sum_{i=1}^n T_i Z_i - n_A \sum_{i=1}^n Z_i}{n_A(n - n_A)}\Bigg| \mathcal{F}_n\right\} \\
&= 0
\end{aligned}
$$

and

$$
\begin{aligned}
\text{Var}\{\bar{Z}_A - \bar{Z}_B\} &= E\text{Var}\left\{\frac{n\sum_{i=1}^n T_i Z_i - n_A \sum_{i=1}^n Z_i}{n_A(n - n_A)}\Bigg| \mathcal{F}_n\right\} \\
&= \frac{n^2 n_A \sigma^2 + n n_A^2 \sigma^2 - 2n n_A^2 \sigma^2}{n_A^2(n - n_A)^2} \\
&= \frac{n\sigma^2}{n_A(n - n_A)} \\
&= \frac{\sigma^2}{Q(1 - Q)n}.
\end{aligned}
$$

Then by an application of Chebyshev's inequality, for any $\epsilon > 0$,

$$
\begin{aligned}
\Pr\{|\bar{Z}_A - \bar{Z}_B| \geq \epsilon\} &\leq \frac{\sigma^2}{\epsilon^2 Q(1 - Q)n} \\
&\to 0 \text{ as } n \to \infty.
\end{aligned}
$$

Note that we have had to assume fixed allocation proportions, which do not arise from complete randomization. Fixed allocation proportions can be obtained from a restricted randomization procedure such as the random allocation rule. The Chebyshev bound does not work under complete randomization, *i.e.*, $T_1, ..., T_n \sim$ i.i.d. $b(p), p \in (0, 1)$ (see Problem 5.1.).

5.3   ACCIDENTAL BIAS

The development in Section 5.2 focuses on the probability of a covariate imbalance, but does not address the effect that such an imbalance may have on the results of the clinical trial. Efron (1971) introduced the term *accidental bias* to describe a measure of the bias in the treatment effect induced by an unobserved covariate. Consider the normal error linear model, from which we will estimate the treatment effect by the ordinary least squares method. Here we modify notation from Chapter 3 slightly. Let $T = (T_1, ..., T_n)'$ be centered treatment indicators, *i.e.*, $T_i = 1$ if treatment A and $T_i = -1$ if treatment B, $i = 1, ..., n$. We will assume that $E(T) = 0$. (Note that all the randomization procedures in Chapter 3 have this property.) Let $Y = (Y_1, ..., Y_n)'$ be a vector of responses to treatment. Suppose we fit a standard normal error regression model, where the mean response, conditional on $T = t = (t_1, ..., t_n)'$, is given by

$$E(Y) = \mu e + \alpha t, \tag{5.1}$$

where $e = (1, 1, ..., 1)'$. Under (5.1), the design matrix is

$$X = [e : t],$$

and hence

$$(X'X)^{-1} = \frac{1}{n^2 - (e't)^2} \begin{bmatrix} n & -e't \\ -e't & n \end{bmatrix},$$

$$X'Y = \begin{bmatrix} e'Y \\ t'Y \end{bmatrix}.$$

Then

$$\hat{\alpha} = \frac{nt'Y - (e't)(e'Y)}{n^2 - (e't)^2}.$$

However, we have ignored a covariate, $z = (z_1, ..., z_n)'$, that is important in the model. Without loss of generality, assume $z'e = 0$ and $z'z = 1$. Then the correctly specified model is given by

$$E(Y) = \mu e + \alpha t + \beta z. \tag{5.2}$$

Taking the expectation with respect to Y when model (5.2) is correct, we obtain

$$\begin{aligned} E(\hat{\alpha}) &= \frac{n(\mu e't + n\alpha + \beta z't) - e't(n\mu + \alpha e't)}{n^2 - (e't)^2} \\ &= \alpha + \frac{n}{n^2 - (e't)^2}\beta z't. \end{aligned}$$

The squared bias term is then given by

$$\{E(\hat{\alpha} - \alpha)\}^2 = \left(\frac{n}{n^2 - (e't)^2}\right)^2 \beta^2 (z't)^2.$$

It is clear that we should desire $e't = 0$, or that the treatment assignments be balanced to minimize accidental bias. If that is accomplished, the degree to which we are subject to accidental bias is controlled by the term $(z't)^2$, which is zero if z is orthogonal to t. Since t is a realization of T, we can obtain the unconditional expectation, by taking $E(z'T)^2$ for a fixed vector z, and we obtain

$$E(z'T)^2 = z'\Sigma_T z,$$

where $\Sigma_T = \text{Var}(T)$. By a result of Rao (1973, p. 62), $z'\Sigma_T z$ cannot exceed the maximum eigenvalue of Σ_T, and the inequality is sharp if the corresponding eigenvector is orthogonal to e. So Efron uses the maximum eigenvalue of Σ_T as a criterion to define the degree to which a randomization procedure is subject to accidental bias. This yields a minimax criterion when used as the basis for determining a randomization procedure $T_1, ..., T_n$ that minimizes the maximum possible value of $z'\Sigma_T z$.

5.4   MAXIMUM EIGENVALUE OF Σ_T

Note that, for complete randomization, $\Sigma_T = I$, and hence the maximum eigenvalue is one. This is the smallest possible value for the maximum eigenvalue of Σ_T (see Problem 5.2). For restricted randomization, it is usually not a trivial exercise to derive the variance-covariance structure of the treatment assignments. In Chapter 3, we were able to derive the covariances for the random allocation rule and the truncated binomial design. We will now explore the behavior of the maximum eigenvalue, denoted λ_{max}, for these two designs.

For the random allocation rule, we have from (3.4) that $\text{cov}(T_i, T_j) = -1/(n-1)$ (the factor 4 in the denominator disappears when the treatment assignments are 1 and -1), so that Σ_T is of the form $aI + bJ$, where $a = 1 + 1/(n-1)$ and $b = -1/(n-1)$. Then we can derive

$$\lambda_{max} = 1 + \frac{1}{n-1}.$$

So as $n \to \infty$, the accidental bias becomes negligibly small compared to complete randomization. From this result we also obtain the maximum eigenvalue for the random allocation rule within a permuted block of size $m = n/M$ as

$$\lambda_{max} = 1 + \frac{1}{m-1}$$

(Problem 5.3).

The truncated binomial design is far more complicated. From (3.10), we have $\text{cov}(T_i, T_j) = \Pr(\tau < i)$, and the probability statement involves the truncated negative binomial distribution in (3.11). For n even, define $H(k) = \sum_{l=1}^{k} \Pr(\tau = n/2 + l - 1)$. Since $\text{cov}(T_i, T_j) = 0$ if $i < (n + 2)/2$, we can write the variance-covariance matrix in a block structure:

$$\Sigma_T = \begin{bmatrix} I & 0 \\ 0 & C \end{bmatrix},$$

where C is an $n/2 \times n/2$ matrix with elements $(1 - \delta_{ij})H(\min(i,j)) + \delta_{ij}$ (δ_{ij} is the Kronecker delta). Rosenberger and Rukhin (2002) then prove that

$$\sqrt{\pi n/3} \leq \lambda_{max} \leq \sqrt{n/2},$$

so that λ_{max} grows like $n^{1/2}$. It is the determinism of the tail sequence that induces correlations that drive the accidental bias to infinity as $n \to \infty$. Consequently, the truncated binomial design is not protective against accidental bias.

5.5 ACCIDENTAL BIAS FOR THE BIASED COIN DESIGNS

The computation of the exact variance-covariance structure for Efron's (1971) biased coin design and its generalizations is not feasible. For his biased coin design, Efron examined the sequence $T_{h+1}, ..., T_{h+n}$, where h and n tend to infinity and computed the autocovariance function

$$\rho_k = \lim_{h \to \infty} \text{cov}(T_h, T_{h+k}).$$

From graphical evidence, he conjectured that the maximum eigenvalue of this long-range variance-covariance structure is $\lambda_{max} = 1 + (2p - 1)^2$. Steele (1980) later proved the result formally. When $p = 2/3$, for example, the degree of accidental bias is $10/9$, equivalent to that of the permuted block design with $m = 10$.

Smith (1984) shows that Efron's solution may be unsatisfactory when there are short-range dependencies in the data. Consider the case where the covariate vector is given by $z_1 = 2^{-1/2}, z_2 = -2^{-1/2}$, and $z_3, ..., z_n = 0$. Then, for the biased coin design,

$$z'\Sigma_T z = 2p \geq 1 + (2p - 1)^2 \tag{5.3}$$

(Problem 5.4). Hence, certain choices of z can behave far worse than Efron's solution ignoring short-range dependencies would suggest.

Smith performs a formal spectral analysis of the generalized biased coin design, given in (3.19), assuming that $z_1, ..., z_n$ form a weakly stationary process. He concludes that the vulnerability to accidental bias is of the order

$$1 + \rho(1 + \rho)(1 + 2\rho)^{-1}n^{-1}\ln n + O(n^{-1}). \tag{5.4}$$

When $\rho = 1$, we have Wei's urn design, and (5.4) reduces to

$$1 + \frac{2}{3}\frac{\ln n}{n} + O(n^{-1}).$$

These values are tabulated for various values of n in Table 5.1.

Table 5.1 *Accidental bias for Wei's urn design (UD), ignoring terms of* $O(n^{-1})$, *for various values of* n.

n	Bias
25	1.09
50	1.05
100	1.03
200	1.02
400	1.00

5.6 SIMULATION RESULTS

The above model, first proposed by Efron, provides theoretical results which describe the maximum susceptibility of a randomization procedure to accidental covariate imbalances. However, this is a "worst case" model, which may not describe the relative properties of each randomization procedure in practice. To explore the practical susceptibility to covariate imbalances, we simulated the probability of a covariate imbalance for various randomization procedures.

For $n = 100$, let $Z_1, ..., Z_n$ be covariate values with mean $\bar{Z}_A$ on treatment A and $\bar{Z}_B$ on treatment B. We computed the probabilities

$$\Pr(|\bar{Z}_A - \bar{Z}_b| > \epsilon)$$

for $\epsilon = 0.5$ and 1.0, where $Z_1, ..., Z_n$ are (1) independently and identically distributed as $N(0,1)$; (2) subject to a drift over time, ranging from -2 to 2 plus a $N(0,1)$ random variable; and (3) autocorrelated. Under model (1), the standard error of the mean difference is 0.20 with perfect balance, larger with an imbalance. Each simulation involved 10,000 replications. For randomization sequences, we used complete randomization (CR), random allocation rule (RAR), truncated binomial design (TBD), permuted blocks with $m = 10$ using the random allocation rule within blocks (PB-RAR) and the truncated binomial design within blocks (PB-TBD), Efron's biased coin design with $p = 2/3$ (BCD), and Wei's urn design $UD(0,1)$ (UD). Results are shown in Table 5.2.

It is clear that for a simple stream of independent and identically distributed continuous covariate values, there is very little probability of a covariate imbalance, as seen theoretically in Section 5.2. However, the probability increases dramatically when we have a drift in covariate values or autocorrelation. In either case, the truncated binomial design results in the most drastic covariate imbalances, particularly when there is a linear drift, with nearly three times the probability of a covariate imbalance with $\epsilon = 0.5$. This matches the theory well. When there is a covariate drift, one can see that blocking minimizes the probability of an imbalance, while the biased coin is nearly as good. Wei's urn design is not quite as good as the biased coin design, but

Table 5.2   Simulated probabilities of a covariate imbalance for three different types of co-variate streams, $n = 100$, 10,000 replications.

Randomization	$\Pr(|\bar{Z}_A - \bar{Z}_B| > \epsilon)$	
	$\epsilon = 0.5$	$\epsilon = 1.0$
$Z_1, ..., Z_n \sim$ i.i.d. $N(0,1)$		
CR	0.0123	0.0000
RAR	0.0115	0.0000
TBD	0.0109	0.0000
PB-RAR	0.0129	0.0000
PB-TBD	0.0117	0.0000
BCD	0.0109	0.0000
UD	0.0119	0.0000
Covariate model with drift		
CR	0.1020	0.0011
RAR	0.1090	0.0010
TBD	0.2880	0.0269
PB-RAR	0.0129	0.0000
PB-TBD	0.0124	0.0000
BCD	0.0264	0.0001
UD	0.0563	0.0002
Covariate model with autocorrelation		
CR	0.0795	0.0005
RAR	0.0790	0.0005
TBD	0.0890	0.0007
PB-RAR	0.0642	0.0004
PB-TBD	0.0804	0.0005
BCD	0.0652	0.0004
UD	0.0704	0.0003

the differences are negligible. We can also note that the probability of a covariate imbalance for the permuted block design are nearly identical whether the random allocation rule or the truncated binomial design were used.

5.7   CONCLUSIONS

While accidental imbalances on covariates, either known or unknown, is a concern, Efron's model describes the potential for severe imbalances. Since a restricted randomization procedure aimed at achieving balance increases the likelihood of periodicity in the sequence of assignments, it also increases the likelihood of periodicity in the sequence of covariate values which resonates with the treatment assignments, resulting in a covariate imbalance. Such situations, however, are extremely rare. In practice it is unlikely that there is a substantial difference in the likelihood of covariate imbalances with any of the restricted procedures considered herein when the sequence of covariate values can be viewed as being drawn from some homogeneous population, perhaps with drift or autocorrelation. The one exception is the truncated binomial design which can result in severe imbalances.

Whether to force balance on known covariates using some form of covariate-adaptive randomization (as discussed in Chapter 4) or to allow the randomization procedure to achieve balancing on its own has been a source of contention among clinical trialists. With respect to a non-randomized covariate-adaptive procedure such as Taves's minimization method, Rosenbaum (1995) takes the view that using such a design to balance on known covariates does not ensure that other unmeasured variables would be similarly balanced. Aickin (2001) takes the view that such nonrandomized covariate-adaptive procedures are acceptable because (i) as noted by Taves (1974), the procedure still incorporates a stochastic element, since the treatment assignments are random variables determined by the stochastic process of incoming covariates; (ii) simulation studies show that covariate-adaptive procedures can improve balance on unknown covariates as well as known covariates; and (iii) selection bias is not an issue with non-randomized studies provided the investigators are masked to the allocation procedure and the allocation sequence is produced at a central core facility.

Our approach is that randomization should be used whenever possible, and covariate-adaptive randomization, such as the Pocock-Simon procedure, are an attractive alternative to Taves's minimization method. However, with respect to randomized covariate-adaptive procedures, to our knowledge, there has been little research done on the effect of such procedures on the balancing of unknown covariates or their susceptibility to selection bias.

To conclude, accidental bias does not appear to be a serious problem for any of the restricted randomization procedures discussed in Chapter 3, with the exception of the truncated binomial design. Similar balancing results can be achieved using the random allocation rule, permuted block design, Efron's biased coin design with $p = 2/3$, or Wei's urn design.

5.8   PROBLEMS

5.1 As in Section 5.2, assume that $Z_1, ..., Z_n$ are independent and identically distributed random variables with mean μ and variance σ^2. Let $T_1, ..., T_n$ be independent and identically distributed Bernoulli random variables with parameter $p \in (0, 1)$. Use Chebyshev's inequality to show that

$$\Pr\left\{|\bar{Z}_A - \bar{Z}_B| \geq \epsilon\right\} \leq \frac{\sigma^2}{\epsilon^2} E\left(\frac{1}{\sum_{i=1}^n T_i} + \frac{1}{\sum_{i=1}^n (1 - T_i)}\right),$$

which does not tend to zero.

5.2 Show that 1 is the smallest possible value for the maximum eigenvalue of Σ_T when $T_j = 1$ or -1.

5.3 a. Show that the variance-covariance matrix for treatment allocation in permuted block randomization using the random allocation rule (within a block i of size $m = n/M$) is given by a block diagonal matrix with diagonal elements

$$\Sigma_{T,i} = \left(1 + \frac{1}{m - 1}\right) I - \frac{1}{m - 1} J, \ i = 1, ..., M,$$

where I is the identity matrix and J is a matrix of 1's.
b. Show that the maximum eigenvalue of Σ_T is given by

$$\lambda = 1 + \frac{1}{m - 1}.$$

c. Graph the vulnerability to accidental bias versus values of m.

5.4 Prove equation (5.3).

5.9 REFERENCES

AICKIN, M. (2001). Randomization, balance, and the validity and efficiency of design-adaptive allocation methods. *Journal of Statistical Planning and Inference* **94** 97–119.

BERRY, D. A. (1989). Comment: ethics and ECMO. *Statistical Science* **4** 306–310.

EFRON, B. (1971). Forcing a sequential experiment to be balanced. *Biometrika* **58** 403–417.

RAO, C. R. (1973). *Linear Statistical Inference*. Wiley, New York.

ROSENBAUM, P. R. (1995). *Observational Studies*. Springer, New York.

ROSENBERGER, W. F. AND RUKHIN, A. L. (2002). Bias properties and non-parametric inference for truncated binomial randomization. Submitted.

SMITH, R. L. (1984). Sequential treatment allocation using biased coin designs. *Journal of the Royal Statistical Society B* **46** 519–543.

STEELE, J. M. (1980). Efron's conjecture on vulnerability to bias in a method for balancing sequential trials. *Biometrika* **67** 503–504.

TAVES, D. R. (1974). Minimization: a new method of assigning patients to treatment and control groups. *Clinical Pharmacology and Therapeutics* **15** 443–453.

6

Selection Bias

6.1 INTRODUCTION

Selection bias refers to biases that are introduced into an unmasked study because an investigator may be able to guess the treatment assignment of future patients based on knowing the treatments assigned to the past patients. Patients usually enter a trial sequentially over time. Staggered entry allows the possibility for a study investigator to alter the composition of the groups by attempting to guess which treatment will be assigned next. Based on whichever treatment is guessed to be assigned next, the investigator can then choose the next patient scheduled for randomization to be one whom the investigator considers to be better suited for that treatment. One of the principal concerns in an unmasked study is that a study investigator might attempt to "beat the randomization" and recruit patients in a manner such that each patient is assigned to whichever treatment group the investigator feels is best suited to that individual patient.

This type of guessing in an unmasked trial could introduce a bias in the composition of the treatment groups which in turn could bias the study results. In principle, randomization is employed to mitigate this bias, and indeed this is a compelling reason to have randomized clinical trials. But the investigator may still be able to guess future treatment assignments with high probability, depending on the randomization procedure employed. As noted in Chapter 3, selection bias arises most frequently in the context of permuted block designs with fixed block sizes, as some treatment assignments will necessarily be deterministic to ensure balance within each block. The unmasking of the sequence of past treatment assignments can allow accurate prediction of future treatment assignments in the same block.

Classic scenarios for selection bias are part of the clinical trialist's folklore. An investigator for a pharmaceutical company, very anxious to see the company's latest pharmaceutical product succeed, guesses the randomization sequence and randomizes patients he or she deems more likely to respond positively to the new therapy when he believes the new therapy to be next in the sequence. A sympathetic nurse coordinator tries to assign a favorite patient to the new therapy rather than placebo. These scenarios are likely to be more the result of subconscious preferences than deliberate dishonesty. Also, any deliberate guessing is likely to be inaccurate, as randomization procedures can be complicated and the investigator may not understand fully the subtleties of the particular procedure used. However, as Smith (1984) points out, they do not have to be right all the time: it is sufficient that they make more right guesses than wrong guesses. While multi-center trials may make it difficult to determine what is going on in other centers, stratification within clinical center eliminates this protection.

The great clinical trialist Chalmers (1990) was convinced that the elimination of selection bias is the most essential requirement for a good clinical trial. He was especially concerned that there are too many loopholes in eligibility criteria and in the rejection of patients during the screening phase, during which the physician could project his or her doubts to the patients while seeking consent. Even when the randomization sequences are intended to be masked, it is not unusual for patients to be unmasked during the course of the trial, due to either adverse events known to be highly associated with one of the treatments, life-threatening emergencies requiring unmasking, or distinguishing features of the masked treatment, such as taste. Regardless of how it arises, selection bias results in inflated type I error rates (Proschan, 1994; Berger and Exner, 1999).

In this chapter, we examine a simple model for susceptibility to selection bias, developed by Blackwell and Hodges (1957), for randomization procedures that are intended to promote balance, such as the random allocation rule and Wei's urn design. We then adjust their model slightly, using Smith's (1984) suggestion, to explore the probability of selection bias for generalized biased coin designs. The model assumes that the investigator will guess the randomization sequence and attempt to put each patient on the treatment that he or she believes is better for that patient.

6.2 THE BLACKWELL-HODGES MODEL

The Blackwell-Hodges model for selection bias assumes that random treatment assignment is independent of the patient characteristics and responses, meaning that an adaptive procedure is not employed. Suppose the primary outcome of a trial is a random variable Y, and the null hypothesis is true; that is, $E(Y|A) = E(Y|B) = \mu$. The experimenter wishes to bias the study by selecting a patient with a higher value of $E(Y)$ when he guesses that A is the next treatment, designated as the guess a, and a lower value of $E(Y)$ when he guesses that B is the next treatment, designated as the guess b. Let $E(Y|a) = \mu + \Delta$ be the expected value of the response when the experimenter guesses a and let $E(Y|b) = \mu - \Delta$ be the expected value of the

Random Assignment

		A	B	Expected Value
Experimenter's	a	$\alpha n/2$	$(1-\beta)n/2$	$\mu + \Delta$
Guess	b	$(1-\alpha)n/2$	$\beta n/2$	$\mu - \Delta$
		$n/2$	$n/2$	

Fig. 6.1   The Blackwell-Hodges model for selection bias.

response when the experimenter guesses b. Of course, the experimenter's guess may be wrong. The accuracy of guessing is described by the parameters (α, β), which represent the probabilities of correct guesses of treatment A and B, respectively. For illustration, assume that n is even, and the number assigned to each treatment is fixed at $n/2$ (the model will also apply when the limiting proportions are $1/2$). Then the expected numbers of guesses, and the expected values of each, are represented in Figure 6.1.

Let G be the total number of correct guesses, and let $\bar{Y}_A$ and $\bar{Y}_B$ be the treatment group means among those randomized to A and B, respectively. Then from Figure 6.1, the expected number of correct guesses is $E(G) = (\alpha + \beta)n/2$. The possible bias introduced by correct guesses is represented by the expected treatment group means among those randomized to each treatment. These are

$$E(\bar{Y}_A) = \frac{(\alpha n/2)(\mu + \Delta) + [(1-\alpha)n/2](\mu - \Delta)}{n/2} = 2\alpha\Delta + (\mu - \Delta),$$

$$E(\bar{Y}_B) = \frac{[(1-\beta)n/2](\mu + \Delta) + [\beta n/2](\mu - \Delta)}{n/2} = -2\beta\Delta + (\mu + \Delta).$$

Then the expected treatment difference is given by

$$E(\bar{Y}_A - \bar{Y}_B) = 2\Delta(\alpha + \beta - 1) = 2\Delta \frac{E(G - n/2)}{n/2}. \qquad (6.1)$$

In (6.1), the investigator's bias 2Δ is the quantity introduced by attempts to beat the randomization. If $\Delta = 0$, we have no bias, and hence the investigator cannot bias the study since there is truly no differential treatment effect between the groups guessed to be assigned to A and the groups guessed to be assigned to B. The *expected bias factor* is the remaining term

$$E(F) = E(G - n/2).$$

If the experimenter guesses completely at random, then $\alpha = \beta = 1/2$ and $E(F) = 0$. These results also apply to unbalanced randomization (Problem 6.1).

Blackwell and Hodges (1957) show that the optimal strategy for the experimenter upon randomizing the jth patient is to guess treatment A when $N_A(j-1) < N_B(j-1)$ and guess treatment B when $N_A(j - 1) > N_B(j - 1)$. When there is a tie, the experimenter guesses with equal probability. Blackwell and Hodges call this the *convergence strategy*.

We see from (6.1) that the expected bias factor $E(F)$ for a randomization procedure where the experimenter employs the convergence strategy can be obtained as follows. For the jth patient, the experimenter guesses treatment A when $N_A(j-1) < N_B(j-1)$ and guess treatment B when $N_A(j-1) > N_B(j-1)$. Among these guesses, now call a correct guess a *hit* and an incorrect guess a *miss*. If $N_A(j-1) = N_B(j-1)$ (*i.e.* we have a *tie*), then the investigator has no basis for a guess and arbitrarily chooses either A or B systematically. Let H, M, and T denote the total number of hits, misses, and ties, in an n-patient randomization stream, respectively. For ties, by chance, $T/2$ are expected to be guessed correctly. Therefore, from the above,

$$E(G) = (\alpha + \beta)n/2 = E[H + T/2].$$

Since $n = E(H + M + T)$, we have

$$E(F) = E(G) - n/2 = E(H - M)/2. \tag{6.2}$$

Equation (6.2) allows us to assess the expected bias factor for any given sequence of random assignments.

For any double-masked randomization, regardless of the method of treatment assignment, since N_A and N_B are unknown to the investigator, the expected number of correct guesses $E(G)$ is simply $n/2$, in which case $E(F) = 0$. The issue, however, is the potential for selection bias in an unmasked study. With complete randomization, $E(F) = 0$ because there is a fixed probability of $1/2$ of assignment to A for all allocations, and thus future random assignments are not in any way predictable based on past assignments. Thus, complete randomization eliminates the expected potential for selection bias. However, this is not the case with restricted randomization procedures which are designed to eliminate or reduce the probability of treatment imbalances. Such sequences are to some degree predictable, and thus are subject to selection bias. In the following sections, the precise expressions for $E(F)$ are presented for various restricted randomization designs.

Stigler (1969) (see also Wei, 1978a) describes the Blackwell-Hodges model in terms of a minimax strategy: they wish to find a design that minimizes the maximum risk, as given by $E(F)$. Stigler proposes that, rather than bias the experiment by Δ on each trial, assume that the investigator picks a subject with expected response between $\mu - \Delta$ and $\mu + \Delta$. Thus the investigator may choose not to bias the experiment at all (selecting a subject with expected response μ), or, at worst, choose according to the Blackwell-Hodges model. Stigler's justification for this more conservative model is that investigators will tend to be timid in realistic situations. He offers the *proportional convergence strategy* to express this timidity. If the investigator observes, after i trials, j treatment A assignments and $i - j$ treatment B assignments, he will then select a subject with expected response

$$\mu + 2\Delta \left(\frac{n/2 - j}{n - i} - \frac{1}{2} \right). \tag{6.3}$$

6.3   SELECTION BIAS FOR THE RANDOM ALLOCATION RULE

We can show that, for the random allocation rule, when the convergence strategy is employed,

$$E(F) = \frac{2^{n-1}}{\binom{n}{n/2}} - \frac{1}{2},$$

(6.4)

as follows.

Think of the random allocation rule as a random walk on a plane starting at point $(0,0)$ and terminating at $(n/2, n/2)$, moving one unit to the right when A is chosen and one unit up when B is chosen. When the walk hits the diagonal, $N_A = N_B$, and the experimenter guesses at random. Away from the diagonal, when the convergence strategy is employed, the experimenter always guesses that the walk moves toward the diagonal. Since the walk begins and ends on the diagonal, it follows that the walk moves toward the diagonal exactly $n/2$ times. Consequently, the experimenter is right at least $n/2$ times. In addition, the experimenter is right, on average, half the time the walk is on the diagonal. Let T be the number of ties. Then

$$E(G) = \frac{n}{2} + \frac{E(T)}{2}.$$

It remains to find $E(T)$. The distribution of T was given by Feller (1950), (but apparently not in later editions):

$$\Pr(T = t) = \frac{2^t \left[\binom{n-t-2}{n/2-t} - \binom{n-t-2}{n/2-t-2} \right]}{\binom{n}{n/2}}.$$

(6.5)

Using (6.5), one can derive

$$E(T) = \frac{2^n}{\binom{n}{n/2}} - 1,$$

(6.6)

(Problem 6.2), and (6.4) follows immediately.

6.4   SELECTION BIAS FOR THE TRUNCATED BINOMIAL DESIGN

The truncated binomial design of Chapter 3 was proposed by Blackwell and Hodges as an alternative to the random allocation rule that would provide less susceptibility to selection bias. For this design, the optimal strategy is to guess the same treatment until $n/2$ of one treatment have been assigned. Then the experimenter should switch to the treatment arm that has less than $n/2$ since all future assignments are known with certainty. We can use the distribution of the number of deterministic selections in the tail, X, given (3.5), to determine $E(F)$. Since the total number of correct guesses under the truncated binomial design is $G = (n - X)/2 + X$, we therefore have

$$E(G) = \frac{n}{2} + \frac{E(X)}{2}. \tag{6.7}$$

Substituting equation (3.6) into (6.7) gives the result

$$E(F) = \frac{n}{2^{n+1}} \binom{n}{n/2}. \tag{6.8}$$

It turns out that, while the results of Chapter 5 show that the truncated binomial design has a high degree of accidental bias, it always has a smaller value of $E(F)$ than the random allocation rule, which we now demonstrate (although the random allocation rule does better than the truncated binomial design when used under Stigler's (1969) proportional convergence strategy in (6.3)). By (6.4) and (6.8), we must show that

$$\frac{n}{2^{n+1}} \binom{n}{n/2} \leq \frac{2^{n-1}}{\binom{n}{n/2}} - \frac{1}{2}$$

or, equivalently,

$$n \binom{n}{n/2}^2 + 2^n \binom{n}{n/2} - 2^{2n} \leq 0, \tag{6.9}$$

for $n \geq 0$, even. For $n = 2$, one can see that the inequality is sharp. Assume (6.9) holds. Then we must show

$$(n + 2) \binom{n+2}{(n+2)/2}^2 + 2^{n+2} \binom{n+2}{(n+2)/2} - 2^{2n+4} \leq 0. \tag{6.10}$$

Noting that

$$\binom{n+2}{(n+2)/2} = 4 \left(\frac{n+1}{n+2} \right) \binom{n}{n/2},$$

the left-hand side of (6.10) is equal to

$$16 \frac{(n+1)^2}{n+2} \binom{n}{n/2}^2 + 16 \left(\frac{n+1}{n+2} \right) 2^n \binom{n}{n/2} - 16 \times 2^{2n}$$

$$= 16 \left[n \binom{n}{n/2}^2 + 2^n \binom{n}{n/2} - 2^{2n} \right] + \frac{16}{n+2} \binom{n}{n/2} \left[\binom{n}{n/2} - 2^n \right].$$

The first term is ≤ 0 by (6.9). For the second term,

$$\binom{n}{n/2} \leq 2^n$$

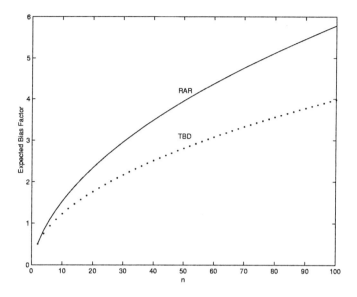

Fig. 6.2 Expected bias factor for the random allocation rule (RAR) and the truncated binomial design (TBD) across values of n.

by a similar induction argument.

Figure 6.2 plots the expected bias factor for the random allocation rule (6.4) and
the truncated binomial design (6.8).

6.5 SELECTION BIAS IN A PERMUTED BLOCK DESIGN

It also follows from the above result that the permuted block design will have less
potential for selection bias when allocations are made using a truncated binomial
than when using a random allocation rule. The explicit relationships follow.

6.5.1 Permuted blocks using the random allocation rule

Under the original Blackwell-Hodges model, each of the M blocks in a permuted-
block design has a potential selection bias equal to that of a random allocation rule
of the same size. Thus, from (6.4), the expected bias factor, for a permuted-block
design with M blocks of equal size $m = n/M$ is

$$E(F) = M \left(\frac{2^{m-1}}{\binom{m}{m/2}} - \frac{1}{2} \right). \tag{6.11}$$

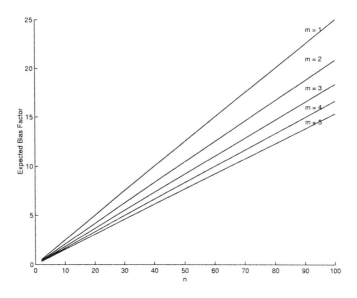

Fig. 6.3   Expected bias factor for the permuted block design with block sizes $m = 2, 4, 6, 8, 10$ for across values of n.

Comparing (6.11) to (6.4), it follows that the expected bias factor for a permuted block randomization with $M > 1$ is always greater than that for the random allocation rule (where $M = 1$). For example, if $n = 100$, then for a random allocation rule $E(F) = 5.78$, whereas for a permuted-block design with five blocks of size 20, $E(F) = 11.69$.

Figure 6.3 shows the total expected bias factor for increasing n for various block sizes of $m = 2, 4, 6, 8$ or 10.

6.5.2   Variable block design

One strategy which has been widely used in an effort to reduce the potential for selection bias with the permuted-block design is to employ a variable block design with random block sizes. Unfortunately, this strategy still yields a substantial potential for selection bias in an unmasked study that is approximately equal to that associated with the average block size.

In general, for a permuted-block design with possibly unequal block sizes $m_i, i = 1, 2, ..., M$, the overall expected selection bias factor is the sum of the individual block selection biases:

$$E(F) = \sum_{i=1}^{M} \left(\frac{2^{m_i - 1}}{\binom{m_i}{m_i/2}} - \frac{1}{2} \right).$$

Since this result applies under the convergent guessing strategy, it is irrelevant whether the investigator is masked or unmasked to the block sizes, or the sequence of block sizes.

Thus, the use of random block sizes does not decrease or eliminate the potential for selection bias. A design employing multiple block sizes has an expected bias factor equal to the sum of the expected bias factors over all blocks. This will approximately equal the expected bias factor associated with M blocks of average block size. For example, the expected bias factor for a sequence where equiprobable random block sizes of 6 and 10 is approximately the same as that for a common block size of 8.

In order to mitigate selection bias, Berger, Ivanova, and Knoll (2002) describe an alternative to the variable block design. They eliminate certain sequences from a fixed block design using a random allocation rule within blocks. Thinking of the random allocation rule as a random walk on a plane (see Section 6.3), within each block, they establish a bound on the distance the random walk can take from the diagonal. The bound is determined by the maximum block size in the variable block design, were that used. They then eliminate all permutation sequences of $m/2$ As and $m/2$ Bs that exceed this bound, and create a discrete uniform distribution across the restricted set of sequences. Consequently, there is more balance throughout the course of the trial, and less predictability than for a variable block design.

6.5.3 Permuted blocks with truncated binomial randomization

An alternative strategy to lessen the susceptibility to selection bias of a permuted block design is to generate the assignments within each block using a truncated binomial design. For a permuted-block design with block size m_i which is known to the investigator, the expected selection bias due to predictions which can be made with certainty under truncated binomial sampling is the sum of the bias factors for each block, which from (6.8) yields.

$$E(F) = \sum_{i=1}^{M} \left[\frac{m_i}{2^{m_i+1}} \left(\begin{array}{c} m_i \\ m_i/2 \end{array} \right) \right].$$

Figure 6.4 compares the expected bias factor across fixed block sizes and $n = 100$, using the random allocation rule and the truncated binomial rule. One can see that the truncated binomial rule results in smaller expected bias than the random allocation rule when used in permuted blocks.

6.5.4 Conclusions

When using a permuted block design, selection bias can be effectively reduced by employing the truncated binomial randomization procedure within blocks rather than the random allocation rule. However, as pointed out in Chapter 5, the risk of accidental bias is much higher for truncated binomial randomization.

It should be noted that the susceptibility to selection bias under either of these models arises because patients are randomized as they arrive. Therefore, the potential

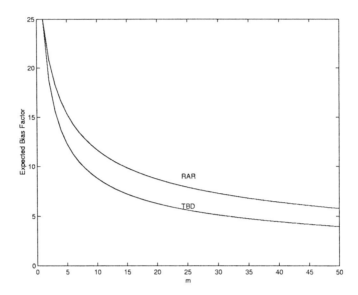

Fig. 6.4 *The effect of block size on the expected bias factor for the permuted block design with* $n = 100$, *comparing the random allocation rule (RAR) and the truncated binomial design TBD.*

for selection bias is completely eliminated if patients are randomized as a block, rather than as they arrive for entry. This is known as *block-simultaneous randomization*. In many trials this will be feasible, especially with small block sizes.

6.6   SELECTION BIAS FOR EFRON'S BIASED COIN DESIGN

Efron (1981) derived the expected selection bias under the convergence strategy. Using the notation in Section 3.6, the probability of a correct guess for the nth allocation is given by

$$P(C_n) = \frac{1}{2}\Pr(|D_{n-1}| = 0) + p\Pr(|D_{n-1}| > 0).$$

As $n \to \infty$, we see that

$$P(C) = \lim_{n \to \infty} P(C_n) = \frac{1}{2}\pi_0 + p(1 - \pi_0) = \frac{1}{2} + \frac{r-1}{4r},$$

by (3.12), where $r = p/(1 - p)$. We can then compute a measure of the asymptotic expected bias factor in n assignments as

$$E(F) = \sum_{i=1}^{n} P(C_i) - n/2 \cong \frac{(r-1)n}{4r}. \tag{6.12}$$

6.7 WEI'S URN DESIGN

For the urn randomization, $UD(\alpha, \beta)$, the probability of assignment to treatment A fluctuates around $1/2$ to a degree proportional to the extent of imbalance. Since $N_a \rightarrow n/2$ as $n \rightarrow \infty$, the degree of fluctuation around $1/2$ decreases as n increases. Thus, future assignments are more predictable early in the sequence of assignments, and the predictability of assignments decreases as n increases. In turn, the potential for selection bias increases initially but then converges to an asymptote as n increases.

Wei (1977) derives the expected bias factor for the urn design. The imbalance after n assignments is $D_n = N_A(n) - N_B(n)$. For any value $|D_n| = d$, $0 \le d \le n$, the probability of a correct guess on the $(n + 1)$th allocation, conditional on $|D_n| = d$, is denoted

$$P(C_{n+1}|d) = \frac{1}{2} + \frac{\beta d}{2(2\alpha + \beta n)}.$$

Therefore, unconditionally, the probability of a correct guess on the $(n + 1)th$ assignment is denoted

$$P(C_{n+1}) = \sum_{d=0}^{n} P(C_n|d) \Pr(|D_n| = d) = 1/2 + \frac{\beta E|D_n|}{2(2\alpha + \beta n)},$$

where $E(|D_n|) = \sum_{d=0}^{n} d \Pr(|D_n| = d)$ can be obtained by the recursive relationship in (3.15). Therefore, the expected bias factor after n assignments for

$$E(F) = \sum_{i=1}^{n} P(C_i) - n/2. \tag{6.13}$$

For comparison, Figure 6.5 presents the expected bias factor for the permuted block design of block size $2m = 10$ using (6.11), the $BCD(2/3)$ (asymptotic) using (6.12), and the $UD(0, 1)$ using (6.13) for $n = 50$ to 100. The urn design has the least potential bias.

6.8 GENERALIZED BIASED COIN DESIGNS

Smith (1984a) considers the limiting value of the selection bias for the class of generalized biased coin designs, given in (3.18), since they do not ensure that exactly $n/2$ patients will be assigned to each treatment. His measure of selection bias is given by

$$\lim_{n \to \infty} \frac{2\Delta E(F)}{n}. \tag{6.14}$$

Under this definition, the random allocation rule and truncated binomial designs each have expected selection bias of order $O(n^{-1/2})$ (Problem 6.3).

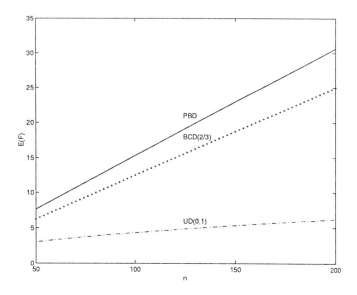

Fig. 6.5 *Expected bias factor for the permuted block design (PBD) with $m = 5$, the biased coin design (BCD) with $p = 2/3$ (asymptotically), and Wei's urn design $UD(0,1)$.*

Using the notation of Section 3.8, the probability of a correct guess for the $(j+1)$th patient, using the convergence criterion, is given by

$$\max \left\{ p\left(\frac{N_A(j) - N_B(j)}{j}\right), 1 - p\left(\frac{N_A(j) - N_B(j)}{j}\right) \right\},$$

and therefore the expected number of correct guesses minus the expected number of incorrect guesses in n patients is

$$\sum_{j=1}^{n-1} \left| 2p\left(\frac{N_A(j) - N_B(j)}{j}\right) - 1 \right| = 2E(G - n/2).$$

Then the expected selection bias is

$$n^{-1} \Delta \sum_{j=1}^{n-1} \left| 2p\left(\frac{N_A(j) - N_B(j)}{j}\right) - 1 \right|. \tag{6.15}$$

Smith (1984b) shows that (6.15) is approximately

$$\Delta\rho\left(\frac{2}{n\pi(1 + 2\rho)}\right)^{1/2}, \tag{6.16}$$

as $n \to \infty$, where ρ is defined in (3.19). As with the random allocation rule and the truncated binomial design, this result is $O(n^{-1/2})$. Using this result, we can directly compare generalized biased coin designs for various values of ρ. For example, when $\rho = 2$, we have approximately 1.55 times the bias of Wei's urn design with $\alpha - 0$ ($\rho = 1$).

6.9   CONTROLLING SELECTION BIAS IN PRACTICE

Berger and Exner (1999) describe several measures that can be taken in a randomized clinical trial to avoid selection bias. Among these are:

1. Maintain a registry of all screened patients, along with a unique identifier, date and time of screening, well-documented rationale for enrollment decisions, and baseline measurements. Having the date and time of screening allows one to determine the treatment group to which the patient would have been enrolled had he or she been randomized.

2. If a treatment code is unmasked for a patient before the completion of enrollment in that patient's block, redefine the block to consist of only those patients enrolled at the time of the unmasking and cease enrollment to the block. Proceed to the next block, possibly appending additional blocks to ensure adequate enrollment.

3. Consider excluding from enrollment decisions investigators who evaluate patients.

4. Do not reuse patient numbers for those patients who have dropped out, and do not bypass new randomization by giving the same treatment to a replacement patient.

We would add that using the truncated binomial design within blocks results in less chance of selection bias than using the random allocation rule, under the Blackwell-Hodges model.

Berger and Exner also suggest testing for unobservable selection bias by examining, within each treatment group, the effect on the response variable of the probability that a patient receives the active treatment. The latter is computed according to the patient's position in the block. This approach complements testing for baseline comparability because it can detect selection bias even when none of the measured baseline variables is imbalanced. If data are collected on patients screened but not randomized, then one can study the joint relationship among baseline covariates, the expected likelihood of a patient to receive the active treatment, and the decision to randomize a patient or not, using regression techniques. If selection bias is detected by these methods, one could then perform a between-group analysis including only those patients for whom selection bias did not compromise the randomization. These patients include those for whom there was complete allocation concealment and the likelihood of receiving the active treatment was 0.5.

6.10   PROBLEMS

6.1 Derive the expected bias factor from the Blackwell-Hodges model when there is fixed unbalanced allocation.

6.2 Derive equation (6.6) from (6.5) (Blackwell and Hodges, 1957).

6.3 Use Stirling's formula to show that $E(F)/n$ for the random allocation rule and truncated binomial design is of order $O(n^{-1/2})$.

6.4 Plot the expected selection bias, $E(F)/n$, versus p for Efron's biased coin design. Is $p = 2/3$ a reasonable choice?

6.11   REFERENCES

BERGER, V. W. AND EXNER, D. V. (1999). Detecting selection bias in randomized clinical trials. *Controlled Clinical Trials* **20** 319–327.

BERGER, V. W., IVANOVA, A., AND KNOLL, M. D. (2002). Enhancing allocation concealment through less restrictive randomization procedures, with application to an unmasked trial of paclitaxel and carboplatin for advanced stage IIIB/IV non-small cell lung cancer. Submitted.

BLACKWELL, D. AND HODGES, J. L. (1957). Design for the control of selection bias. *Annals of Mathematical Statistics* **28** 449–460.

CHALMERS, T. C. (1990). Discussion of biostatistical collaboration in medical research by Jonas H. Ellenberg. *Biometrics* **46** 20–22.

EFRON, B. (1971). Forcing a sequential experiment to be balanced. *Biometrika* **58** 403–417.

FELLER, W. (1950). *An Introduction to Probability Theory and Its Application, Vol. I*. Wiley, New York.

PROSCHAN, M. (1994). Influence of selection bias on type I error rate under random permuted block designs. *Statistica Sinica* **4** 219–231.

SMITH, R. L. (1984a). Sequential treatment allocation using biased coin designs. *Journal of the Royal Statistical Society B* **46** 519–543.

SMITH, R. L. (1984b). Properties of biased coin designs in sequential clinical trials. *Annals of Statistics* **12** 1018–1034.

STIGLER, S. M. (1969). The use of random allocation for the control of selection bias. *Biometrika* **56** 553–560.

WEI, L. J. (1977). A class of designs for sequential clinical trials. *Journal of the American Statistical Association* **72** 382–386.

WEI, L. J. (1978a). On the random allocation design for the control of selection bias in sequential experiments. *Biometrika* **65** 79–84.

WEI, L. J. (1978b). An application of an urn model to the design of sequential controlled clinical trials. *Journal of the American Statistical Association* **73** 559–563.

7

Randomization as a Basis for Inference

7.1 INTRODUCTION

In Chapters 4 through 6 we described several types of bias that can result in biomedical studies, and showed how randomization can mitigate these biases. The second major contribution of randomization is that it can be used as a basis for inference at the conclusion of the trial. Analyses based on a randomization model are completely different from traditional analyses using hypotheses tests of population parameters under the Neyman-Pearson paradigm. In this chapter, we will explore the differences between the randomization model and the population model. In so doing, we will develop the principles of randomization-based inference using permutation tests, originally proposed in the early part of the last century by Fisher (*e.g.*, 1971). A warning to the reader: the Fisher randomization test has contributed to much controversy in the statistical world over recent years. In fact, Fisher himself was somewhat contradictory in his later writings on the subject. For an entertaining and heated debate on the subject, the interested reader is referred to Basu (1980). It should be clear that the authors of this book support randomization-based inference, and the reader should be thus informed. We feel that randomization-based inference is a useful alternative to, or complement to, traditional population model-based methods.

7.2 THE POPULATION MODEL

The most commonly used basis for the development of a statistical test is the concept of a *population model*, where it assumed that the sample of patients is representative of a reference population and that the patient responses to treatment are independent and identically distributed from a distribution dependent on unknown population parameters. In the population model, n_A and n_B patients are randomly sampled from an infinite population of patients on treatment A and treatment B, respectively. Then the $n_i, i = A, B$ patient responses $(Y_{i1}, ..., Y_{in_i})$ can be treated as independent and identically distributed according to some probability distribution $G(y|\theta_i)$ having parameter θ_i. The population model is shown on the left side of Figure 7.1. Under this assumed distribution, it is then a direct matter to construct hypothesis tests comparing the treatment effects, under the Neyman-Pearson lemma, such as

$$H_0 : \theta_A = \theta_B \text{ versus } H_A : \theta_A \neq \theta_B,$$

if θ_i is a scalar. It can also be vector-valued, such as the case where G is normally distributed and $\theta_i = (\mu_i, \sigma^2)$. For this example, the t-test is the uniformly most powerful test of

$$H_0 : \mu_A = \mu_B \text{ versus } H_A : \mu_A \neq \mu_B.$$

Many of the standard statistical tests and estimators based on a population model are developed from the likelihood. We now show that the randomization mechanism is ancillary to the likelihood based on a population model. Let $t^{(j)} = (t_1, ..., t_j)$ and $y^{(j)} = (y_1, ..., y_j)$ be the realized treatment assignments and responses from patients $1, ..., j$, respectively. Let θ be the parameter of interest. Then the likelihood of the data after n patients, denoted $\mathcal{L}_n$, is given by

$$
\begin{aligned}
\mathcal{L}_n &= \mathcal{L}(y^{(n)}, t^{(n)}; \theta) \\
&= \mathcal{L}(y_n | y^{(n-1)}, t^{(n)}; \theta) \mathcal{L}(t_n | y^{(n-1)}, t^{(n-1)}; \theta) \mathcal{L}_{n-1}.
\end{aligned}
\tag{7.1}
$$

Since the responses depend only on the treatment assigned and are independent and identically distributed under a population model, we have

$$\mathcal{L}(y_n | y^{(n-1)}, t^{(n)}; \theta) = \mathcal{L}(y_n | t_n; \theta). \tag{7.2}$$

Also under complete or restricted randomization, the treatment assignments are independent of patient responses, and consequently of θ (this will not be the case for response-adaptive randomization discussed in later chapters). Hence

$$\mathcal{L}(t_n | y^{(n-1)}, t^{(n-1)}; \theta) = \mathcal{L}(t_n | t^{(n-1)}). \tag{7.3}$$

This likelihood reflects the specific restricted randomization procedure employed and the resulting dependence of t_n on $t^{(n-1)}$. Combining (7.1), (7.2), and (7.3), we obtain

$$
\begin{aligned}
\mathcal{L}_n &= \mathcal{L}(y_n | t_n; \theta) \mathcal{L}(t_n | t^{(n-1)}) \mathcal{L}_{n-1} \\
&= \prod_{i=1}^{n} \mathcal{L}(y_i | t_i; \theta) \mathcal{L}(t_i | t^{(i-1)}).
\end{aligned}
\tag{7.4}
$$

Since $\mathcal{L}(t_i|t^{(i-1)})$ is independent of θ, we have

$$\mathcal{L}_n \propto \prod_{i=1}^{n} \mathcal{L}(y_i|t_i; \theta). \qquad (7.5)$$

Note that the likelihood in (7.5) is identical to that arising from a non-randomized design at fixed design points $t_1, ..., t_n$, *i.e.*, for any arbitrary sequence of treatment assignments, including non-random sequences. Consequently, a Bayesian,* or believer in the likelihood principle could use (7.5) to justify an analysis ignoring the randomization mechanism. It is critical to point out that if we followed this approach to inference in this book, we could eliminate all chapters on inference, as the particular randomization procedure used would not matter in our analyses. Any biostatistics textbook would then cover the necessary population-based tests for clinical trials.

Unfortunately, clinical trials do not employ samples of patients that are drawn at random from infinitely large populations of patients on treatment A or treatment B. In fact, there may be no patients at all on treatment A or B to sample from, if the treatments are experimental. Rather, patients are recruited into a clinical trial from various sources by a nonrandom selection of clinics in a nonrandom selection of locations. Clinics are selected because of their expertise, their ability to recruit patients, and their budgetary requirements. From these clinics, a nonrandom selection of eligible and consenting patients is performed, and these patients are then randomized to either treatment A or treatment B.

Nevertheless, it has been argued that these samples of n_A and n_B patients each are, in fact, representative of some larger undefined patient populations, even though they were not truly sampled at random. Arguing in this way, a population model can then be *invoked* as the basis for data analysis, with the assumption that $Y_{ij} \sim G(y|\theta_i)$. The invoked population model is shown on the right side of Figure 7.1. It is important to note that in performing the simplest t-test following a randomized clinical trial, a population model is being invoked.

Even if one could justify an invoked population model, we have discussed only *homogeneous* population models, where each patient is assumed to have the same underlying response distribution, depending only on the treatment assigned. Usually, the characteristics vary according to some underlying characteristics, or vary over time. Therefore, even if patient selection for a trial could be viewed as representative sampling from an unspecified population, often the population would have to be viewed as heterogeneous. In the case of time-heterogeneity, the underlying population model would likely have to incorporate changes over time in some unknown manner.

---

*The Bayesian view on randomization is considerably more complex than this sentence suggests. The role of randomization in Bayesian inference will not be discussed in this book. Some excellent references are Rubin (1978) and Kadane and Seidenfeld (1990).

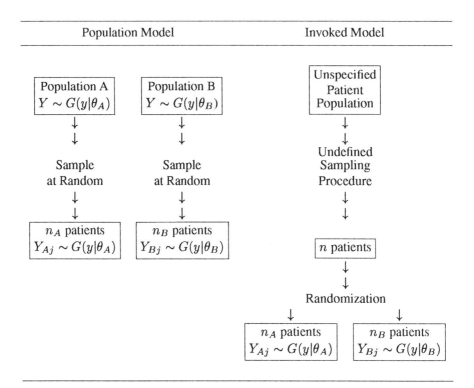

Fig. 7.1 *The population model versus the invoked population model for a clinical trial (Lachin (1988, p. 295), reprinted with permission of Elsevier Science, Inc.).*

In conclusion, as stated by Lachin (1988, p. 296):

> The invocation of a population model for the analysis of a clinical trial becomes a matter of faith that is based upon assumptions that are inherently untestable.

7.3   THE RANDOMIZATION MODEL

As we have seen in Section 7.2, due to the lack of a formal sampling basis, there is no formal statistical foundation for the application of population models to clinical trials. The randomization model is presented in Figure 7.2. Fortunately, the use of randomization provides the basis for an assumption-free statistical test of the equality of the treatments among the n patients actually enrolled and studied. These are known as *permutation tests* or *randomization tests*.

The null hypothesis of a permutation test is that the assignment of treatment A versus B had no effect on the responses of the n patients randomized in the study.

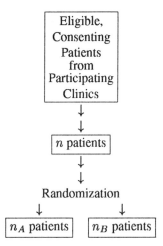

Fig. 7.2 The randomization model for a clinical trial (Lachin (1988, p. 295), reprinted with permission of Elsevier Science, Inc.).

This *randomization null hypothesis* is very different from a null hypothesis under a population model, which is typically based on the equality of parameters from known distributions. The essential feature of a permutation test is that, under the randomization null hypothesis, the set of observed responses is assumed to be a set of deterministic values that are unaffected by treatment. That is, under the null, each patient's observed response is what would have been observed regardless of whether treatment A or B had been assigned. Then the observed difference between the treatment groups depends only on the way in which the n patients were randomized. One then selects an appropriate measure of the treatment group difference, or the treatment effect, which is used as the test statistic. The test statistic is then computed for all possible permutations of the randomization sequence. One then sums the probabilities of those randomization sequences whose test statistic values are at least as extreme as what was observed. This total is then the probability of obtaining a result at least as extreme as the one that was observed, which, by definition, is precisely the p-value of the test. A very small p-value (less than some α, say), then indicates that our observed value is quite extreme compared to the reference set of other possible randomization sequences, and gives strong evidence to conclude that there is a difference between treatments. Permutation tests are assumption-free, but depend explicitly on the particular randomization procedure used.

The simplicity of this approach to inference is often surprising to those rooted in the formal Neyman-Pearson theory of statistical hypothesis testing, and demands some comment. In a sense, it is a direct contradiction to statistical hypothesis testing of a population parameter, because *here we treat the outcome variable of interest as fixed and the treatment assignments (design points) as random; in a population model we traditionally treat the variable of interest as random at fixed values of the design points.* An easy way to think of this is as follows: patients enter the clinical trial with their outcome at the end of the trial pre-stamped on their foreheads, but the outcome is covered. After randomization, the cover is removed and the outcome noted. The only randomness is in the treatment assigned to the patient. If the null hypothesis is true, the outcome values should be evenly distributed across the As and the Bs.

Many statisticians have been convinced of the simple logic behind permutation tests. The following is a quotation attributed to Brillinger, Jones, and Tukey:

> If we are content to ask about the simplest null hypothesis, that our treatment has absolutely no effect in any instance, then the randomization, that must form part of our design, provides the justification for a randomization analysis of our observed result. We need only choose a measure of extremeness of result, and learn enough about the distribution of the result for the observed results held fixed [and] for re-randomizations varying as is permitted by the specification of the designed process of randomization. If p percent of the values obtained by calculating as if a random re-randomization had been made are more extreme than (or equally extreme as) the value associated with the actual randomization, then p percent is an appropriate measure of the unlikeliness of the actual result. Under this very tight hypothesis, this calculation is obviously logically sound. [*Report of the Statistical Task Force to the Weather Modification Advisory Board,* 1978.]

However, a number of questions immediately arise. First, what measure of extremeness, or test statistic, should be used? The most general family of permutation tests is the family of *linear rank tests* (*e.g.,* Lehmann, 1975). Linear rank tests are used often in clinical trials, and the family includes such tests as the traditional Wilcoxon rank-sum test and the logrank test, to name a few. We will focus almost exclusively on linear rank tests in this book.

Second, which set of permutations of the randomization sequence should be used for comparison? If we use all possible permutations, the sequences $AAAA \cdots AA$ and $BBBB \cdots BB$ are included, and these sequences offer no information about the differences between treatments. In fact, if we used a randomization procedure that forces balances between treatments, shouldn't we compare to only those sequences with $n/2$ As and $n/2$ Bs? We will discuss these first two questions in the next two sections.

Third, we have not discussed specific alternative hypotheses or error rates. Without error rates, how can one compute the power of the test? Power can only be determined under an invoked population model. However power and sample size computations under a population model, as shown in Section 2.6, must be considered a crude approximation at best, with measures of variability determined in some sense by "best guesses". It is not unreasonable, therefore, to base sample size computations in planning a study on another "best guess" – an invoked population model, while still relying on a permutation test for analysis.

Fourth, if the analysis of a clinical trial is based on a randomization model that does not in any way involve the notion of a population, how can results of the trial be generalized to determine the best care for future patients? Berger (2000) argues that the difficulty in generalizing to a target population is a weakness not of the permutation test, but of the study design. If it were suspected by investigators that patient experience in a particular clinical trial could not be generalized, there would be no reason to conduct the clinical trial in the first place. Thus we hope that the results of a randomized clinical trial will apply to the general population as well as to the patients in the trial. However, the study design only provides a formal assessment of the latter, not the former. By ensuring validity of the treatment comparison within the trial conducted, by limiting bias and ensuring strict adherence to the protocol, it is more likely that a generalization beyond the trial can be attained.

Lachin (1988) takes the approach that statistical inference in a clinical trial must be viewed as a two-step process. The first step is to determine whether there is a difference between treatments A and B among the n patients actually entered into the trial. The permutation test provides an assumption-free locktight test of this question. The second step is to ascertain the extent to which the observed results can be applied to an invoked population: the hypothetical population from which these n patients arose. For this, it is necessary to invoke a population model. However, this cannot be done with any statistical formalism. Rather, the only recourse is to precisely define the eligibility criteria adopted and then to present distributions of important baseline characteristics in order to describe the hypothetical population from which the study participants arose. The invoked population model then allows the construction of point estimates and confidence intervals and tests of the assumed population parameters.

7.4   PERMUTATION TESTS

The *reference set* of a permutation test is the set of all permutations of randomization sequences that are used to evaluate the tail probability p-value in the comparison with our observed test statistic. An *unconditional reference set* is the set of all possible permutations, including those where all n assignments are to only one treatment A or B, or $n-1$ to only one treatment, etc. This is to be contrasted with a *conditional reference set* which includes only those sequences with the same number of treatments assigned to A and B as were obtained in the particular randomization sequence employed. Let n_A be the observed number of patients assigned to treatment A, *i.e.*, the realization of $N_A(n)$. Let Ω be the cardinality of the reference set, with Ω_u and Ω_c the cardinality of the unconditional and conditional reference sets, respectively. The unconditional reference set will be substantially larger, having

$$\Omega_u = \sum_{n_A=0}^{n} \binom{n}{n_A} = 2^n$$

elements, while the conditional reference set will contain only

$$\Omega_c = \binom{n}{n_A}$$

elements. The conditional reference set is traditionally used because it excludes highly improbable sequences with large imbalances. As we have mentioned, the unconditional reference set contains two sequences (all As and all Bs) that give no information at all about treatment differences and many other sequences with large imbalances that have very little information about treatment differences. Using the conditional reference set is analogous to the traditional argument for conditioning in a population model wherein $N_A(n)$ is an ancillary statistic which provides no information regarding the true treatment difference in the population. In later chapters, when we discuss response-adaptive randomization, $N_A(n)$ is no longer an ancillary statistic, and the same arguments do not apply.

When the random allocation rule or the truncated binomial design is used, there is no distinction between the unconditional and conditional reference sets, as we force $N_A(n) = n_A = n/2$. While the reference sets for the two designs are the same, the sequences in the reference sets have different probabilities. Also, the conditional reference set for complete randomization is equivalent to the reference set for the random allocation rule only on those occasions when we obtain $n_A = n/2$ following complete randomization.

Let S be the test statistic of interest, which can be any measure of the difference between the treatment groups. Define S_l to be the value of S for sequence $l, l = 1, ..., \Omega$ and define $S_{obs.}$ to be our observed test statistic. Let L record realizations of particular randomization sequences; L has a probability distribution depending on the particular randomization procedure employed. Then the p-value of the *unconditional permutation test* is given by

$$p_u = \sum_{l=1}^{\Omega_u} I(S_l \geq S_{obs.}) \Pr(L = l), \tag{7.6}$$

and the *conditional permutation test* is given by

$$p_c = \sum_{l=1}^{\Omega_c} I(S_l \geq S_{obs.}) \Pr(L = l | N_A(n) = n_A), \tag{7.7}$$

where $I(\cdot)$ is the indicator function.

These p-values are one-sided. One rejects the null hypothesis of no difference among the treatments for the n patients studied when $p < \alpha$ for some $\alpha \in (0, 1)$. The logic of including the observed sequence in the reference set has been argued extensively for exact tests, and compromises such as the mid p-value, where only half the probability of the observed sequence is included in the sum, have been suggested. The discreteness of the p-value becomes less relevant as the sample size becomes larger. The reader is referred to Agresti (1990) for more details.

7.5 LINEAR RANK TESTS

A special beauty of the permutation test is that any desired measure of a difference between treatments can be selected, such as difference of means or proportions, etc.

One need never formally describe the distribution or moments of a test statistic, as the randomization-based inference is completely determined by a p-value. However, this presumes that the entire reference set can be enumerated so that the permutation p-value can be computed. This is a tall order, even with today's computing, for even moderate sample sizes (see Section 7.8).

The family of linear rank tests is often used as the basis for a permutation test. Most standard nonparametric tests fall in this family, and these tests tend to have well-defined asymptotic distributions (see Chapter 14), and hence provide a nonparametric method for large-sample inference. Let $Y_1, ..., Y_n$ be the outcomes of the n patients and define a score function a_{jn} to be some score of the jth patient out of n patients, with arithmetic mean $\bar{a}_n$. Let $T_j, j = 1, ..., n$ be 1 if patient j was assigned to A and 0 if B. Then the linear rank statistic is given by

$$S = \sum_{j=1}^{n}(a_{jn} - \bar{a}_n)T_j. \tag{7.8}$$

The particular linear rank test is determined by the choice of a score function. For example, if the $\{a_{jn}\}$ are simple ranks, the linear rank statistic is the well-known Wilcoxon rank-sum test.

As an illustration, consider a clinical trial of four treatments with observed sequence $ABBA$. Suppose the patient outcomes were $Y_1 = 3$, $Y_2 = 1$, $Y_3 = 4$, $Y_4 = 5$. Then the simple ranks are $\{2, 1, 3, 4\}$. Under complete randomization, Table 7.1 gives a complete enumeration of the unconditional and conditional reference sets and the associated values of the Wilcoxon rank-sum test. From (7.6), we see that $p_u = 1/4$ and from (7.7), we see that $p_c = 1/3$.

Now suppose we were using Wei's $UD(0, 1)$ design. Now the sequences are not equiprobable, and the respective probabilities are shown in Table 7.2. Here we compute $p_u = 1/4$ and $p_c = 1/4$.

For binary response data, one can assign binary scores $a_{jn} = 1$ or 0. Under a population model, the resulting linear rank test is algebraically equivalent to the usual Mantel-Haenszel chi-square test for the 2×2 contingency table under a conditional complete randomization model (see Section 8.2).

For survival data, the logrank test can be obtained using *Savage scores* (Kalbfleisch and Prentice, 1980). In the usual notation of survival analysis, $\tau_1, ..., \tau_n$ are the event times of patients $1, ..., n$, and in the simplest case of no ties or censoring, we have n distinct ordered survival times $\tau_{(1)}, ..., \tau_{(n)}$ corresponding to treatment assignments $T_{(1)}, ..., T_{(n)}$. Then the linear rank statistic can be written as

$$S_n = \sum_{j=1}^{n} a_{(j)n} T_{(j)},$$

where

$$a_{(j)n} = 1 - \sum_{k=n+j-1}^{n} \frac{1}{k}, \tag{7.9}$$

Table 7.1   *Unconditional and conditional reference sets for computation of the linear rank test from complete randomization (Lachin (1988, p. 298), reprinted with permission of Elsevier Science, Inc.).*

Unconditional ($\Omega_u = 16$)			Conditional ($\Omega_c = 6$)		
Sequence (*l*)	$\Pr(L = l)$	S_l	Sequence (*l*)	$\Pr(L = l)$	S_l
$AAAA$	1/16	0.0	$AABB$	1/6	−2.0
$AAAB$	1/16	−1.5	$ABAB$	1/6	0.0
$AABA$	1/16	−0.5	$ABBA$	1/6	1.0
$AABB$	1/16	−2.0	$BAAB$	1/6	−1.0
$ABAA$	1/16	1.5	$BABA$	1/6	0.0
$ABAB$	1/16	0.0	$BBAA$	1/6	2.0
$ABBA$	1/16	1.0			
$ABBB$	1/16	−0.5			
$BAAA$	1/16	0.5			
$BAAB$	1/16	−1.0			
$BABA$	1/16	0.0			
$BABB$	1/16	−1.5			
$BBAA$	1/16	2.0			
$BBAB$	1/16	0.5			
$BBBA$	1/16	1.5			
$BBBB$	1/16	0.0			

which is equivalent to the logrank statistic. We can also write the scores in (7.9) as

$$a_{jn} = 1 - E(X_{(j)}), \tag{7.10}$$

where $X_{(1)}, ..., X_{(n)}$ are the order statistics from a unit exponential (Prentice, 1978; Kalbfleisch and Prentice, 1980, p. 79).

With censored data, let $C_1, ..., C_n$ be the censoring times and $D_1, ..., D_n$ be the event times of patients $1, ..., n$. For the jth patient, we can only observe data pairs (Y_j, δ_j), where $Y_j = \min(D_j, C_j)$ and $\delta_j = I(D_j \leq C_j)$, where I is the indicator function. Assume that the censoring mechanism is the same in both treatment groups and that there are no ties. Let $\tau_{(1)} < \tau_{(2)} < \cdots < \tau_{(M)}$ denote the M ordered distinct event times with R_m the number of study patients at risk just prior to $\tau_{(m)}$, $m = 1, ..., M$. Then for the patient with an event at $\tau_{(m)}$, $\delta_j = 1$ and $Y_j = \tau_{(m)}$. For a censored patient, $\delta_j = 0$ and $\tau_{(m)} \leq Y_j < \tau_{(m+1)}$. Then for the logrank test, the appropriate scores are given by

$$a_{(j)n} = 1 - \sum_{m=1}^{j} \frac{1}{R_m}$$

Table 7.2   *Unconditional and conditional reference sets for computation of the linear rank test from the $UD(0, 1)$ (Wei and Lachin (1988, p. 352), reprinted with permission of Elsevier Science, Inc.).*

Unconditional ($\Omega_u = 16$)			Conditional ($\Omega_c = 6$)		
Sequence (l)	$\Pr(L = l)$	S_l	Sequence (l)	$\Pr(L = l)$	S_l
$AAAA$	0	0.0	$AABB$	0	−2.0
$AAAB$	0	−1.5	$ABAB$	1/4	0.0
$AABA$	0	−0.5	$ABBA$	1/4	1.0
$AABB$	0	−2.0	$BAAB$	1/4	−1.0
$ABAA$	1/12	1.5	$BABA$	1/4	0.0
$ABAB$	1/6	0.0	$BBAA$	0	2.0
$ABBA$	1/6	1.0			
$ABBB$	1/12	−0.5			
$BAAA$	1/12	0.5			
$BAAB$	1/6	−1.0			
$BABA$	1/6	0.0			
$BABB$	1/12	−1.5			
$BBAA$	0	2.0			
$BBAB$	0	0.5			
$BBBA$	0	1.5			
$BBBB$	0	0.0			

if $\delta_j = 1$ and

$$a_{(j)n} = -\sum_{m=1}^{j} \frac{1}{R_m}$$

if $\delta_j = 0$. If there are tied event times, the scores are calculated as though there were no ties, and then each of the patients with tied times is assigned the average of their scores.

7.6   VARIANCE OF THE LINEAR RANK TEST

The variance of the linear rank test can either be described with respect to the unconditional reference set or the conditional reference set of permutations. The unconditional variance of the linear rank test can be computed directly from (7.8) as

$$
\mathrm{Var}(S) = \sum_{j=1}^{n}(a_{jn} - \bar{a}_n)^2 \mathrm{Var}(T_j)
$$

$$
+ \sum_{j=1}^{n}\sum_{\substack{i=1\\i\neq j}}^{n}(a_{in} - \bar{a}_n)(a_{jn} - \bar{a}_n)\mathrm{cov}(T_i, T_j). \tag{7.11}
$$

For complete randomization and the random allocation rule, $\mathrm{Var}(T_j)$ and $\mathrm{cov}(T_i, T_j)$ do not depend on i or j, and (7.11) reduces to

$$
\mathrm{Var}(S) = \{\mathrm{Var}(T_j) - \mathrm{cov}(T_i, T_j)\} \sum_{j=1}^{n}(a_{jn} - \bar{a}_n)^2. \tag{7.12}
$$

We can compute this quantity directly from $\boldsymbol{\Sigma}_T$. For complete randomization, from (7.12), we have

$$
\frac{1}{4}\sum_{j=1}^{n}(a_{jn} - \bar{a}_n)^2.
$$

For the random allocation rule, using (3.4) and (7.12), we obtain

$$
\mathrm{Var}(S) = \frac{n}{4(n-1)}\sum_{j=1}^{n}(a_{jn} - \bar{a}_n)^2. \tag{7.13}
$$

The truncated binomial design variance is given by

$$
\mathrm{Var(S)} = \frac{1}{4}\sum_{j=1}^{n}(a_{jn} - \bar{a}_n)^2
$$

$$
+ \frac{1}{2}\sum_{i=(n+2)/2}^{n}\mathrm{Pr}(\tau > i)(a_{in} - \bar{a}_n)\sum_{j=i+1}^{n}(a_{jn} - \bar{a}_n); \tag{7.14}
$$

the derivation is left as an exercise. For most other randomization procedures, an exact form of $\mathrm{Var}(S)$ is intractable, such as for Efron's biased coin design and Wei's urn design because the exact form of $\boldsymbol{\Sigma}_T$ is unknown.

The conditional variance with respect to the conditional reference set, is defined as $\mathrm{Var}(S|N_A(n) = n_A)$. Note that $\mathrm{cov}(T_j, T_{j'})$ is no longer 0 for complete randomization. In fact, $T_j, j = 1, ..., n$, are then dependent Bernoulli indicators with parameter n_A/n, so that $\mathrm{Var}(T_j) = n_A n_B/n^2$, where $n_B = n - n_A$. To find $\mathrm{cov}(T_i, T_j)$, we compute, for $j > i$,

$$
\begin{aligned}
E(T_i T_j) &= \mathrm{Pr}(T_i = 1, T_j = 1)\\
&= \mathrm{Pr}(T_j = 1|T_i = 1)\mathrm{Pr}(T_i = 1)\\
&= \frac{n_A - 1}{n - 1}\left(\frac{n_A}{n}\right).
\end{aligned}
$$

Then

$$\text{cov}(T_i, T_j) = \frac{n_A - 1}{n - 1}\left(\frac{n_A}{n}\right) - \frac{n_A^2}{n^2}$$

$$= -\frac{n_A n_B}{n^2(n - 1)}.$$

Finally we compute

$$\text{Var}(S|N_A(n) = n_A) = \frac{n_A n_B}{n}\left(\sum_{j=1}^{n}(a_{jn} - \bar{a}_n)^2 \bigg/ (n - 1)\right). \quad (7.15)$$

Note that (7.15) reduces to (7.13) when $n_A = n/2$.

It is instructive to compare the variance of the linear rank test under a randomization model with the variance that would be obtained under a population model. For in a population model, one would assume that $t_1, ..., t_n$ are deterministic treatment indicators and $A_{1n}, ..., A_{nn}$ are independent and identically distributed random scores. The linear rank test under a population model can then be written as

$$S_{pop.} = \sum_{j=1}^{n} t_j(A_{jn} - \bar{A}_n) = \sum_{j=1}^{n}(t_j - n_A/n)A_{jn}$$

and

$$\text{Var}(S_{pop.}) = \text{Var}(A_{jn})\sum_{j=1}^{n}(t_j - n_A/n)^2$$

$$= \frac{n_A n_B}{n}\text{Var}(A_{jn}). \quad (7.16)$$

We immediately see that the conditional linear rank test for complete randomization under a randomization model has a variance (7.15) that is a consistent estimator of the variance of the linear rank test under a population model, as given in (7.16). This observation gives more insight into the differences between the randomization and population models. If, in fact, we can assume that patient response are independent and identically distributed, then, at least in very large trials using complete randomization or the random allocation rule, the variance of the test statistic will be equivalent under the two models. However, other randomization procedures do not have this property, and as we have said, the assumption of a homogeneous population model may not be appropriate.

7.7   OPTIMAL RANK SCORES

In their classic text, Hájek and Šidák (1967) provide a general approach to the development of a nonparametric test for the comparison of two or more populations. They show the form of the optimal score generating function when it is desired to test the null hypothesis against a location or scale shift when sampling from a specific distribution. The resulting test is optimal in the sense of maximizing the Fisher's information in the data, and thus is asymptotically fully efficient. For example, simple rank (Wilcoxon) scores are optimal to detect a location shift when sampling from a logistic distribution, while van der Waerden scores (see Problem 7.4) are optimal for a normal distribution. Likewise, Savage scores are optimal to detect a scale shift in an exponential distribution. This theory was also used by Peto and Peto (1972) and by Prentice (1978) to derive the optimal rank scores for censored data under a proportional hazards and proportional odds alternative, the logrank and modified Wilcoxon scores, respectively. Thus, there are a wide variety of score functions $\{a_{jn}\}$ that could be employed.

Under the randomization model, the responses are treated as fixed quantities, likewise the rank scores $\{a_{jn}\}$. Since the population model concept of sampling at random from two population distributions does not apply, the concept of efficiency does not strictly apply to the family of linear rank tests with a randomization-based distribution. However, one can still think about the average behavior of the test in repeated similar experiments. In this case the expected properties of the observed responses might be relevant in the choice of the rank scores employed in the analysis. For example, if one thinks that the data from similar experiments are more likely to satisfy a proportional hazards alternative with censored data than a proportional odds alternative, one would choose to employ logrank scores in the analysis rather than modified Wilcoxon scores.

For simple quantitative responses, there is a greater range of choices. Among these, simple rank scores are most commonly employed, in part due to the Mann-Whitney representation of the test, under a population model, as a function of $P(Y_A > Y_B)$ where Y_A represents a random observation from group A and Y_B likewise from group B. The Wilcoxon test statistic provides an estimate of this "proversion" probability, a useful quantity under a population model regardless of the underlying distributions. While the "average" behavior of any one score function over a range of alternatives has not been thoroughly explored, it is reasonable to expect that the simple rank scores will yield a test that is in general robust to a location shift in any distribution.

Of course, the score function to be employed in any analysis must be prespecified. To compute multiple tests using different scores, or to examine the properties of the data to choose the "best" score function would be cheating. Another approach would be to use a score-robust test that provides good power, in a population model sense, over a range of possible alternatives. One such test is the Gastwirth (1966) maximin efficient robust test. This test is a convex combination of the standardized test (Z) values from the "extreme" pair in the set of tests considered. The extreme pair is determined by the estimated asymptotic relative efficiency of each pair of tests, which is equivalent to the square of the correlation of each pair. This approach could also be employed with the families of tests herein.

Consider two different tests using scores a_{jn} and b_{jn}, such as Wilcoxon scores and Savage scores. Let $\boldsymbol{a} = (a_{1n}, ..., a_{nn})'$ and $\boldsymbol{b} = (b_{1n}, ..., b_{nn})'$ refer to the corresponding vectors of scores and let Σ_T refer to the covariance matrix of the vector of treatment assignments $(T_1, ..., T_n)$. Then the covariance of the two test statistics is simply $a'\Sigma_T b$, from which the correlation between the two tests is obtained. Based on the resulting correlations between each pair of tests in the family of tests under consideration, the combination of tests is selected so as to maximize the minimum asymptotic relative efficiency relative to whichever test in the family would actually be optimal, if such were known in advance. See Lachin (2000, Sec. 4.9.2) for the required expressions. This approach, however, is only applicable to those randomization designs for which the Σ_T is known explicitly. This includes complete randomization, the truncated binomial design, the random allocation rule, and the permuted block design, but not Efron's biased coin design or Wei's urn designs, or their generalizations, for which Σ_T is not known.

7.8 CONSTRUCTION OF EXACT PERMUTATION TESTS

Even for conditional tests, enumerating all possible permutations in the reference set becomes prohibitively large as n gets larger than around 15. Mehta, Patel, and Wei (1988) give a computational algorithm which is effective for computing the exact distribution of permutation tests following restricted randomization procedures. Even so, such algorithms are probably only reasonable for sample sizes less than 50 unless parallel processing is used.

The basic networking algorithm is as follows. Let $P_{j+1}(n_A) = E(T_{j+1}|N_A(j) = n_{Aj})$, so that the algorithm applies to all restricted randomization procedures for which the $(j+1)$th treatment assignment depends on the previous treatment assignments only through $N_A(j)$ (this applies to the restricted randomization designs in Chapter 3). Let $T = (T_1, ..., T_n)$ and let $\Omega_{n_A} = \{T : N_A(n) = n_{An}\}$. One does not have to enumerate every sequence in Ω_{n_A} in order to compute the exact distribution of the test statistic. The networking algorithm begins with a single node $(0,0)$. For $j = 1, ..., n-1$, each node (j, n_{Aj}) generates nodes $(j+1, n_{A,j+1})$ ending in a single terminal node (n, n_{An}). To each distinct subpath

$$(0,0) \to (1, n_{A1}) \to \cdots \to (j, n_{Aj}),$$

assign a rank length

$$a_{1n}n_{A1} + a_{2n}(n_{A2} - n_{A1}) + \cdots + a_{jn}(n_{Aj} - n_{A,j-1})$$

with associated probability

$$\prod_{k=1}^{j} \{P_k(n_{A,k-1})\}^{n_{Ak} - n_{A,k-1}} \{1 - P_k(n_{A,k-1})\}^{1-(n_{Ak} - n_{A,k-1})}.$$

Some of the rank lengths will not be unique; suppose there are $l(n_{Aj})$ distinct rank lengths, and denote them as $S_{jl}, l = 1, ..., l(n_{Aj})$. Let π_{jl} be the sum of the probabilities for those paths that have the same rank length. Then the set $\Omega(j, n_{Aj}) = \{(S_{jl}, \pi_{jl}, l = 1, ..., l(n_{Aj}))\}$ is the probability distribution of $S_j = \sum_{i=1}^{j} a_{in}T_i$ given $N_A(j) = n_{Aj}$. One can then obtain each set $\Omega(j+1, n_{A,j+1})$ recursively from

$\Omega(j, n_{Aj})$. The degree of computational efficiency gained by eliminating redundant sequences with the same rank length will of course depend on the scores a_{jn}. For the simple rank scores, this number increases as $O(n^2)$. However, for the logrank test, the algorithm increases exponentially with n. Hollander and Peña (1988) give an algorithm for exact enumeration when there are $K > 2$ treatments using Markov chain techniques.

The *Monte Carlo* approach has become popular in computing permutation tests (see, for example, Good, 1994 and Edgington, 1995). Here a sample of the possible permutations is drawn at random, and this sample is used in place of the complete reference set. The Monte Carlo approach yields approximate p-values, that are largely based on how large and representative the sampling procedure is. Numerical analysis algorithms, such as the *branch and bound algorithm* from combinatoric optimization can also be used to eliminate the need to enumerate each permutation.

There has been little literature on randomization-based inference following covariate-adaptive randomization, such as the Pocock-Simon procedure (Section 4.4.2). Simon (1979, p. 508) advocates a randomization analysis using simulation:

> It is possible, though cumbersome, to perform the appropriate permutation test generated by a nondeterministic adaptive stratification design. One assumes that the patient responses, covariate values, and sequence of patient arrivals are all fixed. One then simulates on a computer the assignment of treatments to patients using the [Pocock-Simon procedure] and the treatment assignment probabilities actually employed. Replication of the simulation generates the approximate null distribution of the test statistic adopted, and the significance level. One need not make the questionable assumption that the seqeuence of patient arrivals is random.

7.9   LARGE SAMPLE PERMUTATION TESTS

While the Monte Carlo and numerical analysis approaches can be implemented fairly quickly with appropriate software, because most phase III clinical trials involve large numbers of patients, the asymptotic distribution of the linear rank test has been most often employed in practice. In fact, large-sample approximations to the linear rank test from complete randomization and Wei's urn design are quite accurate, even for samples as small as $n = 20$ (Mehta, Patel, and Wei, 1988).

In general, one would presume that test statistics of the form

$$W = \frac{S}{\sqrt{\text{Var}(S)}}$$

should follow a standard normal distribution for large samples, based on our knowledge of the central limit theorem. In many cases this will be true, but the theory is complicated because the treatment assignments are correlated under the particular restricted randomization procedure. Chapter 14 gives the necessary theoretical developments.

The main condition for asymptotic normality is a Lindeberg-type condition on the scores, requiring that

$$\lim_{n \to \infty} \frac{\max_{1 \leq j \leq n} (a_{jn} - \bar{a}_n)^2}{\sum_{j=1}^{n} (a_{jn} - \bar{a}_n)^2} \to 0. \tag{7.17}$$

While this condition looks complicated, it essentially says that no individual absolute score can grow too large relative to the sum of all the absolute scores. For example, we could not use the actual data values from a continuous unbounded random variable. It is easy to see, for instance, that the simple ranks satisfy (7.17), as

$$\frac{\max_{1 \leq j \leq n} (a_{jn} - \bar{a}_n)^2}{\sum_{j=1}^{n} (a_{jn} - \bar{a}_n)^2} = \frac{(n-1)^2}{\sum_{j=1}^{n} (j-1)^2} = \frac{6(n-1)}{n(2n-1)} \to 0$$

as $n \to \infty$, at a rate $O(1/n)$. Many other common score functions satisfy (7.17) as well (see Problem 14.1). If the scores satisfy the condition, then the unconditional test

$$W_U = \frac{2 \sum_{j=1}^{n} (a_{jn} - \bar{a}_n) T_j}{\sqrt{\sum_{j=1}^{n} (a_{jn} - \bar{a}_n)^2}} \tag{7.18}$$

will be asymptotically standard normal. In particular, (7.18) is the correct form of the test statistic under complete randomization, the random allocation rule, and Wei's urn design. However, this has not been proved for the truncated binomial design or Efron's biased coin design, and these are still open problems, as is the large sample distribution of the linear rank test following covariate-adaptive randomization.

For the conditional test, conditional on $N_A(n) = n_A$, under complete randomization, the test statistic

$$W_C = \frac{\sum_{j=1}^{n} (a_{jn} - \bar{a}_n) T_j}{\sqrt{\gamma n_A (n - n_A)/n}}, \tag{7.19}$$

has an asymptotic standard normal distribution, where γ is defined by the additional assumption that

$$\gamma = \lim_{n \to \infty} \frac{\sum_{j=1}^{n} (a_{jn} - \bar{a}_{jn})^2}{n}. \tag{7.20}$$

For example, with continuous observations, defining the scores as $a_{jn} = r_{jn}/(n+1)$ where r_{jn} are the simple ranks, (7.20) is satisfied, and we have $\gamma = 12$ if there are no ties (Problem 14.4). Note that the addition to the sum of squares when there are ties is asymptotically negligible, provided the number of ties does not grow with n. However, this could be problem for some outcomes, such as ordinal measures or continuous measures that are truncated to integer values (*e.g.*, age). If there are many ties, one could substitute the observed value of (7.20) for γ in (7.19).

The above tests should be easy to compute in any statistical software package. In SAS, one could use the RANK procedure to compute the score function.

While these tests are simple to compute, under $UD(\alpha, \beta)$ randomization, the form of the conditional test statistic is complicated. Define a sequence of modified scores

$$
\begin{aligned}
b_{nn} &= a_{nn} - \bar{a}_n, \\
b_{jn} &= (a_{jn} - \bar{a}_n) \\
&\quad - \beta(2\alpha + (j-1)\beta) \sum_{k=j+1}^{n} \frac{(a_{kn} - \bar{a}_n)}{(2\alpha + (k-1)\beta)(2\alpha + (k-2)\beta)}, \\
&\qquad j = 1, ..., n-1,
\end{aligned} \tag{7.21}
$$

and let the $\{a_{jn}\}$ sequence be normalized so that $\sum_{j=1}^{n} b_{jn}^2 = 1$. Also define another sequence of modified scores, denoted $\{\tilde{b}_{jn}\}$, which are computed by substituting $a_{jn} - \bar{a}_n = n^{-1/2}$ for all j into (7.21). Define

$$
\rho_n = \sum_{j=1}^{n} b_{jn}\tilde{b}_{jn}, \quad s^2 = \lim_{n \to \infty} \sum_{j=1}^{n} \tilde{b}_{jn}^2.
$$

Then, conditional on $D_n = N_A(n) - N_B(n) = d_n$, the test is given by

$$
W_C = \frac{2\sum_{j=1}^{n}(a_{jn} - \bar{a}_n)T_j - 2\rho_n d_n/(\sqrt{n}s^2)}{\sqrt{1 - \rho_n^2/s^2}}.
$$

While it may be tempting to employ the form of the test in (7.19), in this case simulations show that the test is slightly anti-conservative, and the more complicated form is more appropriate. SAS code to compute this test is given in Appendix B of this chapter. The program accepts binary indicator variables to designate treatment group, but the code is based on expressions developed in Chapter 14.

The asymptotic form of the conditional test for Efron's biased coin design is an open problem.

Table 7.3 gives results of a sample data analysis of cholesterol data from the Diabetes Control and Complications Trial on 50 patients. The data can be found in Table 7.4 in Appendix A of this chapter. Since these are "baseline" data, the null hypothesis applies. We also generate a single pass simulation of 50 treatment assignments, under complete randomization, the random allocation rule, and Wei's $UD(0, 1)$. Using the simple rank scores, we compute the values of the linear rank statistic and associated p-values in Table 7.3. We see that the values of the tests are quite different, depending on the particular randomization procedure used and whether the conditional or unconditional test is employed.

7.10   GROUP SEQUENTIAL MONITORING

In many clinical trials, it is desirable to establish a sequential monitoring plan, whereby the test statistic is computed at an interim point or points in the trial and a

Table 7.3 Values of the test statistic and p-values for the linear rank test (simple rank scores) for the data in Table 7.4.

Randomization Procedure	Test Value	p-value (2-sided)
Complete unconditional	-0.510	0.610
Complete conditional	-0.503	0.614
Random allocation	0.265	0.791
$UD(0, 1)$ unconditional	0.098	0.922
$UD(0, 1)$ conditional	0.118	0.906

decision is made whether to stop early due to evidence of treatment efficacy, while preserving the overall type I error rate. When the test statistic is computed and decisions are made after groups of patients have responded to treatment, such a plan is called *group sequential monitoring*. There is a large literature on group sequential monitoring of population-based inference procedures; see Jennison and Turnbull (2000) for a comprehensive overview of the subject. We are unaware of literature on group sequential monitoring of permutation tests. Here, consider just a single interim monitoring point after n_1 patients have responded; the basic formulation can be extended to any number of inspections. Let

$$S_{n_1} = \sum_{j=1}^{n_1}(a_{jn_1} - \bar{a}_{n_1})T_j = \boldsymbol{a}'_{n_1}\boldsymbol{T}_{n_1}$$

be the computed linear rank statistic after n_1 patients and let

$$S_n = \sum_{j=1}^{n}(a_{jn} - \bar{a}_n)T_j = \boldsymbol{a}'_n\boldsymbol{T}_n$$

be the computed statistic at the end of the trial, where $\boldsymbol{a}'_{n_1} = (a_{1n_1} - \bar{a}_{n_1}, ..., a_{n_1n_1} - \bar{a}_{n_1})$, $\boldsymbol{a}'_n = (a_{1n} - \bar{a}_n, ..., a_{nn} - \bar{a}_n)$, $\boldsymbol{T}'_{n_1} = (T_1, ..., T_{n_1})$, and $\boldsymbol{T}'_n = (T_1, ..., T_n)$.

It is necessary to find the joint probability distribution of (S_{n_1}, S_n). This could be computed exactly, as in Lin, Wei, and DeMets (1991) or using the asymptotic joint distribution. Using the approach of Slud and Wei (1982), we are interested in finding constants c_1 and c_2 such that

$$\Pr(S_{n_1} > c_1) = \alpha_1$$

and

$$\Pr(S_n > c_2, S_{n_1} \leq c_1) = \alpha_2, \tag{7.22}$$

where $\alpha_1 + \alpha_2 = \alpha$ and α is the overall desired size of the test. Alternatively, the *spending function approach* of Lan and DeMets (1983) requires specifying a strictly increasing continuous function $\alpha^*(t), t \in [0, 1]$ such that $\alpha^*(0) = 0$ and $\alpha^*(1) = \alpha$. Then we find constants d_1 and d_2 such that

$$\Pr(S_{n_1} > d_1) = \alpha^*(t_1)$$

and

$$\Pr(S_n > d_2, S_{n_1} \leq d_1) = \alpha^*(1) - \alpha^*(t_1), \tag{7.23}$$

where t_1 represents the fraction of information available, with respect to total information accrued in the entire clinical trial, after n_1 patients.

In the usual case where the test statistic is a sum of independent and identically distributed random variables, we could write $S_n = S_{n_1} + S_{n_2}$, where S_{n_1} and S_{n_2} are independent. This simplifies the distribution theory dramatically. Unfortunately for the linear rank test, the first n_1 elements of $\{a_{jn}\}$ are not necessarily the same as $\{a_{jn_1}\}$, and hence the linear rank statistic cannot be decomposed. Even if it could be decomposed in this way, S_{n_1} and S_{n_2} would not be independent, except under complete randomization.

If unconditional inference is used, we can easily compute the covariance of the test statistics as follows. Let $\mathbf{\Sigma}_{n_1} = \text{Var}(\mathbf{T}_{n_1})$ and $\mathbf{\Sigma}_n = \text{Var}(\mathbf{T}_n)$ Assuming that (S_{n_1}, S_n) are jointly asymptotically normal and $\mathbf{\Sigma}_n$ does not depend on n, we can see that $\text{Var}(S_{n_1}) = \mathbf{a}'_{n_1} \mathbf{\Sigma}_{n_1} \mathbf{a}_{n_1}$, $\text{Var}(S_n) = \mathbf{a}'_n \mathbf{\Sigma}_n \mathbf{a}_n$ and

$$\text{cov}(S_{n_1}, S_n) = [\mathbf{a}'_{n_1} : \mathbf{0}_{1 \times n - n_1}] \mathbf{\Sigma}_n \mathbf{a}_n. \tag{7.24}$$

The probabilities in (7.22) and (7.23) can be computed using this covariance under the correct asymptotic joint distribution.

For conditional inference, the variance-covariance structure of $T_1, ..., T_{n_1}, \mathbf{\Sigma}_{n_1}$, will be determined conditional on $N_A(n_1) = n_{A1}$, say, and this will no longer be a submatrix of $\mathbf{\Sigma}_n$, since it will be a function of n_{A1}. In this case, the variance-covariance matrix of treatment assignments among the first n_1 allocations with the vector of n allocations is more complicated than for the unconditional test. Note that this approach is applicable only to those designs for which the unconditional variance-covariance matrix of treatment assignments can be described explicitly: namely, complete randomization, the permuted block design, the random allocation rule, and the truncated binomial design.

While a general procedure for developing the theory for a sequential monitoring strategy is apparent from the above discussion, it should be clearly stated that this problem has not been addressed in the literature. Finding the joint asymptotic distribution of sequentially computed linear rank statistics under different randomization procedures using given score functions is an open topic. From (7.24), it is clear that in many cases the test statistics do not have independent increments.

If we use the Slud and Wei approach, we can select any α_1 and α_2 that satisfies $\alpha_1 + \alpha_2 = \alpha$. One must be careful in its implementation though. It is preferable that the number of interim inspections (K) and the sequence $\{\alpha_1, ..., \alpha_K\}$ be pre-specified, so that the selection of α_j will not depend on the data.

The spending function approach allows for any arbitrary sequence of interim inspections, and unplanned inspections can be added by simply computing $\alpha^*(t)$ for the given information at the additional interim inspection. For the spending function approach, we must determine the fraction of total information accrued in the trial after n_1 patients. Under a population model, the information attained by n_1 would be defined according to the Fisher's information for the estimator of the parameter of interest, which is asymptotically equivalent to the inverse of the variance of the estimator (to first order). For an estimation-based test, constructed as the ratio of the estimator to its standard error, then the variance of the test is decreasing in n, and information increasing. However, for such tests expressed as a partial sum rather than a mean, the expression for the information is proportional to the variance of the sum, with both increasing in n.

Since the linear rank test involves the sum of the scores, this suggests that the variance of the test could be used as a measure of the information in the data. Thus, even though we are not operating under a population model, the information fraction at n_1 is given by

$$ t_1 = \frac{a'_{n_1} \Sigma_{n_1} a_{n_1}}{a'_n \Sigma_n a_n}. $$

In general, while Σ_n will be known for some randomization procedures, we will not know a_n. Thus it is necessary to employ a surrogate measure of information, as in Lan and Lachin (1990). Let $n_A(t_1)$ and $n_B(t_1)$ be the number of patients assigned to treatment A and B, respectively, at the time of the interim inspection and let $n(t_1) = n_A(t_1) + n_B(t_1)$. Since the trial will have a target sample size in each group, n_A and n_B, where $n = n_A + n_B$, then one possible surrogate could be

$$ \tilde{t} = \frac{\frac{n_A(t_1)n_B(t_1)}{n(t_1)}}{\frac{n_A n_B}{n}} = \frac{q_1(1 - q_1)n_1}{q(1 - q)n}, $$

where $q_1 = n_A(t_1)/n(t_1)$ and $q = n_A/n$. This is the information fraction from the usual t-test. Then the spending function can be computed with respect to the $\tilde{t}$ values. See Lan and Lachin (1990) for more details on the appropriateness of using surrogate measures of information when the true information fraction cannot be computed.

7.11   PROBLEMS

7.1 Verify equations (7.1) and (7.4).

7.2 Read Basu (1980) and the ensuing discussion (including that of the venerable discussant, the late Oscar Kempthorne). Prepare a five minute position paper expressing your views on permutation tests. Present your paper in a classroom debate with fellow students. Focus on the following issues:

(i) Did Fisher contradict himself or change positions on the role of permutation tests?
(ii) Does Basu make a convincing argument in his example of the scientist and the

statistician?

(iii) Is there a middle ground?

7.3 For the examples in Table 7.1 and 7.2, compute the following:

a. the unconditional and conditional p-values if Efron's biased coin design had been used with $p = 2/3$;

b. the p-value if the truncated binomial design had been used (see Table 3.2);

c. the conditional p-value for the $UD(0, 1)$ if the actual randomization sequence had been $ABAA$ instead of $ABBA$.

7.4 Let r_{jn} be the simple rank scores. The *van der Waarden scores* are defined as

$$a_{jn} = \Phi^{-1}\left(\frac{r_{jn}}{n+1}\right)$$

(Lehmann (1975, p. 97)), where Φ is the standard normal distribution function. Recompute the example in Table 7.1 using the van der Waarden scores instead of simple rank scores.

7.5 Verify equation (7.12).

7.6 Derive the variance in (7.14).

7.7 Verify (7.24).

7.9 Redo the data analysis in Table 7.3 using the van der Waarden scores (Problem 7.4) for cholesterol values in Table 7.4.

7.12 REFERENCES

AGRESTI, A. (1990). *Categorical Data Analysis*. Wiley, New York.

BASU, D. (1980). Randomization analysis of experimental data: the Fisher randomization test. *Journal of the American Statistical Association* **75** 575–595, with discussion.

BERGER, V. W. (2000). Pros and cons of permutation tests in clinical trials. *Statistics in Medicine* **19** 1319–1328.

EDGINGTON, E. S. (1995). *Randomization Tests*. Marcel Dekker, New York.

FISHER, R. A. (1935). *The Design of Experiments*. Oliver and Boyd, Edinburgh.

GASTWIRTH, J. L. (1966). On robust procedures. *Journal of the American Statistical Association* **61** 929–948.

GOOD, P. (1994). *Permutation Tests*. Springer, New York.

HÁJEK, J. AND ŠIDÁK, Z. (1967). *Theory of Rank Tests*. Academia, Prague.

HOLLANDER, M. AND PEÑA, E. (1988). Nonparametric tests under restricted treatment-assignment rules. *Journal of the American Statistical Association* **83** 1144–1151.

JENNISON, C. AND TURNBULL, B. W. (2000). *Group Sequential Methods with Applications to Clinical Trials*. Chapman and Hall/CRC, Boca Raton.

KADANE, J. B. AND SEIDENFELD, T. (1990). Randomization in a Bayesian perspective. *Journal of Statistical Planning and Inference* **25** 329–345.

KALBFLEISCH, J. D. AND PRENTICE, R. L. (1980). *The Statistical Analysis of Failure Time Data.* Wiley, New York.

LACHIN, J. M. (1988). Statistical properties of randomization in clinical trials. *Controlled Clinical Trials* **9** 289–311.

LACHIN, J. M. (2001). *Biostatistical Methods: The Assessment of Relative Risks* . Wiley, New York.

LAN, K. K. G. AND DEMETS, D. L. (1983). Discrete sequential boundaries for clinical trial. *Biometrika* **70** 659–663.

LAN, K. K. G. AND LACHIN, J. M. (1990). Implementation of group sequential logrank tests in a maximum duration trial. *Biometrics* **46** 759–770.

LEHMANN, E. L. (1986). *Nonparametrics: Statistical Methods Based on Ranks.* Holden-Day, San Francisco.

LIN, D. Y., WEI, L. J., AND DEMETS, D. L. (1991). Exact statistical inference for group sequential trials. *Biometrics* **47** 1399–1408.

MEHTA, C. R., PATEL, N. R., AND WEI, L. J. (1988). Constructing exact significance tests with restricted randomization rules. *Biometrika* **75** 295–302.

PETO, R. AND PETO, J. (1972). Asymptotically efficient rank invariant test procedures. *Journal of the Royal Statistical Society A* **135** 185–206, with discussion.

PRENTICE, R. L. (1978). Linear rank tests with right-censored data. *Biometrika* **65** 167–179.

RUBIN, D. B. (1978). Bayesian inference for causal effects: the role of randomization. *Annals of Statistics* **6** 34–58.

SIMON, R. (1979). Restricted randomization designs in clinical trials. *Biometrics* **35** 503–512.

SLUD, E. AND WEI, L. J. (1982). Two sample repeated significance tests based on the modified Wilcoxon statistic. *Journal of the American Statistical Association* **77** 862–868.

WEI, L. J. AND LACHIN, J. M. (1988). Properties of urn randomization in clinical trials. *Controlled Clinical Trials* **9** 345–364.

7.13 APPENDIX A: DCCT DATA

Table 7.4 gives data from the Diabetes Control and Complications Trial used in the data analysis example in Section 7.9.

Table 7.4 *Cholesterol levels from 50 patients and simulated randomization sequences under complete randomization, random allocation rule (RAR), truncated binomial (TBD), and $UD(0,1)$.*

Patient	Cholesterol	Complete	RAR	$UD(0,1)$
1	132	1	1	1
2	195	1	1	0
3	157	0	1	1
4	196	1	0	0
5	190	0	1	1
6	228	1	0	0
7	191	1	0	0
8	150	1	1	1
9	154	1	0	0
10	147	0	0	0
11	207	1	0	0
12	113	0	0	1
13	174	1	1	1
14	210	0	1	1
15	144	1	1	1
16	217	1	1	1
17	167	0	0	0
18	229	0	0	0
19	123	1	1	1
20	248	1	0	0
21	146	0	0	0
22	193	0	0	0
23	182	0	1	1
24	116	1	1	1
25	189	0	1	1
26	215	0	1	1
27	211	1	1	1
28	252	0	1	1
29	206	0	1	1
30	151	1	0	0
31	232	1	1	1
32	238	1	0	0
33	201	0	0	0
34	174	0	0	0
35	151	0	0	0

Patient	Cholesterol	Complete	RAR	$UD(0,1)$
36	150	1	0	0
37	221	0	1	1
38	232	0	1	1
39	179	1	1	1
40	167	1	0	0
41	213	1	0	0
42	153	1	0	0
43	122	1	0	0
44	204	1	0	0
45	196	1	1	1
46	168	0	1	1
47	196	0	1	1
48	185	0	0	0
49	235	1	0	0
50	143	1	1	0

7.14   APPENDIX B: SAS CODE FOR CONDITIONAL $UD(0,1)$ LINEAR RANK TEST

Here we provide SAS code to compute the asymptotic conditional linear rank test for $UD(0,1)$ randomization, based on data in Table 7.4.

```
*Enter data - $j$ is patient number;
data dcct;
 input j cholest treat;
 treat=2*treat-1;
 cards;
1 132  1
2 195  0
3 157  1
4 196  0
 .
 .
 .
;

*Find simple ranks;
proc rank data=dcct out=a;
 ranks simrank;
 var cholest;
*Compute mean of ranks;
proc means data=a noprint;
 var simrank;
 output out=b mean=meanrank n=numobs;
*Compute $D_n$;
```

```
proc means data=a noprint;
 var treat;
 output out=c sum=dn;
*Compute centered ranks;
data two;
 set a;
 if _N_=1 then set b;
 if _N_=1 then set c;
 cenrank=simrank-meanrank;
 keep j cenrank treat dn numobs;
*Find the scaling factor;
data three;
 set two;
 if j > 2 then term=cenrank/((j-1)*(j-2));
 else term=0;
proc sort; by descending j;
proc iml;
 use three;
 read all var {j term} into x;
 n = nrow(x);
 sum = j(n,1,0);
 sum[1] = x[1,2];
 sum[2:n] = cusum(x[1:n-1,2]);
 x = x||sum;
 varn={j term cusum};
 create data from x(|colname=varn|);
 append from x; close data;
 quit;
data four;
 merge three data; by descending j;
 bjn=cenrank-(j-1)*lag(cusum);
 if bjn=. then bjn=cenrank;
 bjnsq=bjn**2;
proc means noprint data=four;
 var bjnsq;
 output out=five sum=bjnsqsum;
data six;
 set four;
 if _N_=1 then set five;
 scaled=cenrank/sqrt(bjnsqsum);
 if j > 2 then newterm=scaled/((j-1)*(j-2));
 else newterm=0;
 keep j scaled treat newterm dn numobs;
*Compute $b_{jn}$'s and $\tilde{b}_{jn}$'s, $s$, and $x$;
 proc iml;
```

```
 use six;
 read all var {j newterm} into x;
 n = nrow(x);
 sum = j(n,1,0);
 sum[1] = x[1,2];
 sum[2:n] = cusum(x[1:n-1,2]);
 x = x||sum;
 varn={j newterm cusum};
 create newdata from x(|colname=varn|);
 append from x;
 close newdata;
 quit;
data seven;
 merge six newdata; by descending j;
 bjn=scaled-(j-1)*lag(cusum);
 if bjn=. then bjn=scaled;
 bjntilde=(j-1)/((numobs-1)*sqrt(numobs));
 prod=bjn*bjntilde;
 num=scaled*treat;
 ssq=1/3;
 x=dn/sqrt(numobs);
*Compute $\rho$ and $S_n$;
proc means data=seven noprint;
 var prod num;
 output out=rho sum=sumprod sn;
*Compute the test statistic;
data eight;
 set seven;
 if _N_=1 then set rho;
 rhosq=sumprod**2;
 testnum=sn-(sumprod*x/ssq);
 testden=sqrt(1-(rhosq/ssq));
 test=2*testnum/testden;
 pvalue=2*(1-probnorm(abs(test)));
data nine;
 set eight;
 if _N_=1;
proc print data=nine;
 var test pvalue;
```

8

Inference for Stratified, Blocked, and Covariate-Adjusted Analyses

8.1 INTRODUCTION

As described in Chapter 4, in many studies a stratified or blocked analysis is desired. Under a randomization model, the proper analysis for any stratified or blocked randomization is a *like-stratified analysis*. Methods for stratified analysis of proportions, or means, or lifetables have been derived under population model sampling, such as the Mantel-Haenszel test for a stratified analysis of 2×2 tables. These methods can also be applied under a randomization model, using the appropriate randomization model variances.

It is tempting, however, to avoid the computational complications of such an analysis in favor of the simpler unstratified analysis, an analysis that is inherently simpler to describe. Such an unstratified analysis can be justified under a homogeneous population model, meaning that within all strata or blocks all subjects were drawn at random from a single homogeneous population. Of course this assumption may be untenable in a multi-center clinical trial stratified by clinic, or in trials stratified by gender or age, as common examples. One question, therefore, is whether any loss is incurred whenever the stratified/blocked analysis is appropriate, but the unstratified analysis is performed. Thus we also examine the difference between a stratified versus an unstratified analysis under both a population and a randomization model.

In some studies it is also desired to conduct an analysis that is "adjusted" for other covariates. Such an adjusted analysis can be performed using post-stratification, meaning stratification by a factor that was not also employed in the pre-stratification of the randomization. Alternately a regression model could be used to describe the

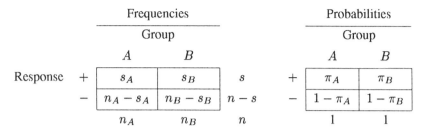

Fig. 8.1 *Notation for the Mantel-Haenszel procedure for two-group comparisons in a 2 × 2 contingency table.*

association between post-hoc covariates and to compute covariate-adjusted values, and covariate-adjusted rank scores. Examples of both approaches are presented.

8.2   STRATIFIED ANALYSIS

8.2.1   The Mantel-Haenszel procedure

Perhaps the simplest, and most common, instance of a stratified analysis is the Mantel-Haenszel test for multiple independent 2×2 tables (Mantel and Haenszel, 1959). The principle of the Mantel-Haenszel stratified test also generalizes to other settings, such as the analysis of variance (*cf.* Fleiss, 1986). We first describe the test for a single 2×2 table under a population model, and then under a randomization model. We then generalize the test to the case of I 2×2 tables.

Within a single 2×2 table, the frequencies and the corresponding probabilities under a population model can be expressed in Figure 8.1. Under a population model, n_A, n_B, and n are fixed. Additionally, if we condition on the total number of responses $S = s$ (*cf.* Lachin, 2000), then under the null hypothesis $H_0 : \pi_A = \pi_B$, the only randomness in the table is the upper left cell, the number of responses in group A, S_A, with realization s_A. Then S_A is distributed as a central hypergeometric distribution with expectation

$$E(S_A) = n_A s/n \tag{8.1}$$

and variance

$$\text{Var}(S_A) = \frac{n_A n_B s(n-s)}{n^2(n-1)}. \tag{8.2}$$

Under H_0, asymptotically for large n,

$$S_A - E(S_A) \sim N\left(0, \text{Var}(S_A)\right),$$

and thus

$$X_U^2 = \frac{(S_A - E(S_A))^2}{\text{Var}(S_A)} \tag{8.3}$$

is asymptotically distributed as χ^2 on 1 degree of freedom, under H_0.

This same test can also be derived as a special case of the linear rank test under a simple randomization model. Assume that the treatment assignments are generated by complete randomization and that we condition on the numbers assigned to each group, which are now random ($N_A = n_A$, $N_B = n_B$). Then the randomization distribution of the linear rank test with binary scores can be derived. This test can be expressed only in terms of S_A, the number of events on treatment A, and this is asymptotically normally distributed with expectation and variance that are identical to the expressions (8.1) and (8.2) obtained under a population model. Equivalent expressions are also obtained for the conditional randomization distribution based on a random allocation rule for which the total sample sizes are fixed and equal such that $n_A = n_B = n/2$. Further, S_A, suitably normalized, is asymptotically normally distributed under a randomization model with mean $s/2$ and variance $s(n - s)/4(n - 1)$.

Now consider the case of I independent strata or blocks. Within the ith stratum of size n_i ($i = 1, \dots, I$), the 2×2 table contains positive frequencies $\{s_{iA}, s_{iB}\}$ with marginal totals $n_{iA}, n_{iB}, s_i, n_i - s_i$. Under a population model, again conditioning on both margins fixed (cf. Lachin, 2000), under the null hypothesis H_{0i}: $\pi_{iA} = \pi_{iB}$, the number of responses on A within the ith stratum S_{iA} is distributed as a central hypergeometric distribution with expectation

$$E(S_{iA}) = n_{iA} s_i / n_i \tag{8.4}$$

and variance

$$\text{Var}(S_{iA}) = \frac{n_{iA} n_{iB} s_i (n_i - s_i)}{n_i^2 (n_i - 1)}. \tag{8.5}$$

Under H_{0i} asymptotically for large n_i within the ith stratum,

$$S_{iA} - E(S_{iA}) \sim N(0, \text{Var}(S_{iA})).$$

Since the strata are independent, then for fixed I, under the joint null hypothesis H_{0i}: $\pi_{iA} = \pi_{iB}$, $i = 1, \dots, I$,

$$\sum_{i=1}^{I} (S_{iA} - E(S_{iA})) \sim N\left(0, \sum_{i=1}^{I} \text{Var}(S_{iA})\right).$$

Since this is the sum of asymptotically normally distributed stratum-specific variates, then the stratified-adjusted Mantel-Haenszel test is

$$X_S^2 = \frac{\left\{\sum_{i=1}^{I} (S_{iA} - E(S_{iA}))\right\}^2}{\sum_{i=1}^{I} \text{Var}(S_{iA})}, \tag{8.6}$$

and asymptotically X_S^2 is distributed as χ^2 on 1 degree of freedom as $n_i \to \infty$ for all i. The asymptotic distribution can also be demonstrated for the case where the sample size within each stratum is small but the number of strata increases indefinitely (Breslow, 1981).

This stratified-adjusted test can also be derived under a simple randomization model. Within each stratum, assume that the treatment assignments are generated by complete randomization and that we condition on the numbers assigned to each group N_{iA}, N_{iB} as before. Then under the null hypotheses H_{0i} within the ith stratum, the randomization distribution of the number of events in the first group can be derived and S_{iA} is asymptotically normal with expectation and variance (8.4) and (8.5) as obtained under a population model.

Equivalently, under a random allocation rule for which the total sample sizes are fixed and equal such that $n_{iA} = n_{iB} = n_i/2$, the randomization distribution of S_{iA} is distributed as central hypergeometric with

$$
\begin{aligned}
E(S_{iA}) &= s_i/2 \\
\mathrm{Var}(S_{iA}) &= \frac{s_i(n_i - s_i)}{4(n_i - 1)},
\end{aligned}
$$

and within each stratum $S_{iA} - E(S_{iA})$ is asymptotically normally distributed.

Therefore, for fixed I, the stratified test X_S^2 in (8.6) is also asymptotically distributed as χ^2 on 1 degree of freedom under either a population model, or a randomization model conditionally following complete randomization, or following a random allocation rule, within each stratum.

The stratified test, therefore, would be appropriate for any study that employed complete randomization or a random allocation rule within each of I strata, or a randomization that employed such assignments within permuted blocks, with fixed or permuted block sizes. This would also apply to a study that employed such permuted blocks within strata, such as clinics, in which case I is the total number of blocks employed in all strata combined. For studies that employ other randomization procedures, such as the urn design, it is more convenient to describe an analysis in terms of the linear rank test of Chapter 7.

8.2.2 Linear rank test

A more general linear rank test can be applied to data on any scale (quantitative, ordinal, nominal) and for survival data as shown in Section 7.5. For the case of a single stratum or block of size n, the linear rank statistic with centered scores is then defined as

$$
W = \frac{\sum_{j=1}^{n}(a_j - \bar{a})T_j}{\left[\mathrm{Var}\sum_{j=1}^{n}(a_j - \bar{a})T_j\right]^{1/2}} = \frac{S}{V^{1/2}},
$$

where $T_j = 1$ or 0, and we have suppressed the dependence of the scores on n for convenience. As noted in Section 7.9, under conditions on the scores for certain randomization procedures, W is asymptotically distributed as standard normal under the null hypothesis of no treatment effect.

The variance of the statistic can either be estimated under a population or a randomization model, as described in Chapter 7. Under a randomization model, the scores $\{a_j\}$ are considered fixed. The random component, therefore, is the sequence of treatment assignments $\{T_j\}$. Under this model the variance expression V depends on the covariance matrix of the treatment assignments.

For binary scores, $a_j = 1$ or 0, it is readily shown that this test using the conditional variance following complete randomization, or the variance for a random allocation rule, is equivalent to the Mantel-Haenszel test X_U^2 in (8.3) for a single 2×2 table (Problem 8.1).

For the random allocation rule, from (7.13), the statistic using the randomization variance is

$$W_{RAR} = \frac{\sum_{j=1}^{n}(a_j - \bar{a})T_j}{\left[\frac{n}{4(n-1)}\sum_{j=1}^{n}(a_j - \bar{a})^2\right]^{1/2}}. \tag{8.7}$$

The extension of the linear rank test for a prospectively stratified randomization within each of I strata is rather straightforward. For the jth patient in the ith stratum of total size n_i, let $T_{j(i)}$ refer to the randomized treatment assignment and $a_{j(i)}$ refer to the rank score, $j = 1, \ldots, n_i, i = 1, \ldots, I$. The scores $a_{j(i)}$ are a function only of the responses among the n_i patients in the ith pre-randomization stratum, and not patients randomized within other strata. Then let S_i be the corresponding test statistic with variance V_i, using stratum-specific mean-centered scores $(a_{j(i)} - \bar{a}_i)$, with stratum-specific mean $\bar{a}_i$. If asymptotically, under the null hypothesis, S_i is normally distributed within each stratum, for large n_i, then any linear combination $W = \sum_i \omega_i S_i$ based on weights $\{\omega_i\}$ is asymptotically normally distributed with mean zero and variance $\sum_i \omega_i^2 V_i$, and therefore,

$$W_S = \frac{\sum_{i=1}^{I}\omega_i S_i}{\left[\sum_{i=1}^{I}\omega_i^2 V_i\right]^{1/2}} = \frac{\sum_{i=1}^{I}\omega_i \sum_{j=1}^{n_i}(a_{j(i)} - \bar{a}_i)T_{j(i)}}{\left[\sum_{i=1}^{I}\omega_i^2 \,\text{Var}\sum_{j=1}^{n_i}(a_{j(i)} - \bar{a}_i)T_{j(i)}\right]^{1/2}}, \tag{8.8}$$

where ω_i is the weight for stratum i.

For the case of a permuted block randomization in a clinical trial with n patients, using a random allocation rule with block size $m = n/M$ within each of M blocks, then the blocked rank test is

$$W_B = \frac{\sum_{i=1}^{M}\omega_i \sum_{j=1}^{m}(a_{j(i)} - \bar{a}_i)T_{j(i)}}{\left[\frac{m}{4(m-1)}\sum_{i=1}^{M}\omega_i^2 \sum_{j=1}^{m}(a_{j(i)} - \bar{a}_i)^2\right]^{1/2}}. \tag{8.9}$$

Based on the permuted-block randomization, W_B is asymptotically distributed as a standard normal under the null hypothesis. To allow for unfilled blocks, the equivalent expression is

$$W_B = \frac{\sum_{i=1}^{M} \omega_i \sum_{j=1}^{m} (a_{j(i)} - \bar{a}_i) T_{j(i)}}{\left[\sum_{i=1}^{M} \frac{n_{iA} n_{iB}}{n_i(n_i-1)} \omega_i^2 \sum_{j=1}^{m} (a_{j(i)} - \bar{a}_i)^2 \right]^{1/2}}. \tag{8.10}$$

Note that in a stratified randomization using permuted blocks, M is the total number of blocks for all strata combined, and the randomization analysis is as above summing over all blocks in all strata. When the block sizes are small the block-specific components individually may not be normally distributed. However, as $M \to \infty$, using the developments in Breslow (1981), the statistic W_B is asymptotically normally distributed.

The choice of weights $\{\omega_i\}$ in the stratified (blocked) test may be arbitrary, or may be based on efficiency considerations derived from a population model. For Wilcoxon scores, van Elteren (1960) showed that the optimal stratum weights are $\omega_i = (n_i + 1)$ in the sense that these weights maximize the asymptotic efficiency of the stratified test under a local alternative with a common shift between groups under a specific population model. Thus, with equal stratum sizes, or block lengths, the ω_i cancel from the numerator and denominator in (8.8). Puri (1965) similarly shows the form of the optimal weights for other rank scores, such as the Savage or van der Waerden scores, under a local alternative.

An appropriate set of weights for any test under a population model can also be derived as follows. Note that the rank statistic S_i is the difference between groups in the partial sums of the scores in the ith stratum. Then, assuming n_{iA} and n_{iB} are fixed and the scores $\{A_j\}$ are random with realizations $\{a_j\}$, it is readily shown that

$$S_i = \frac{n_{iA} n_{iB}}{n_i} (\bar{A}_{iA} - \bar{A}_{iB}),$$

where $\bar{A}_{iA}$ is the mean score within group A in the ith stratum, and $\bar{A}_{iB}$ is the mean score within group B. Thus

$$\bar{A}_{iA} - \bar{A}_{iB} = S_i \frac{n_i}{n_{iA} n_{iB}} \tag{8.11}$$

with variance

$$\text{Var}(\bar{A}_{iA} - \bar{A}_{iB}) = V_i \left[\frac{n_i}{n_{iA} n_{iB}} \right]^2.$$

Under a population model assuming a common difference in mean scores, using weighted least squares, it is well known that the weights inversely proportional to these variances yield a minimum variance linear estimator (MVLE) of the common mean difference. Thus the optimal linear combination of the mean differences is

$$\sum_{i}^{I} \left[\frac{n_{iA} n_{iB}}{n_i} \right]^2 V_i^{-1} (\bar{A}_{iA} - \bar{A}_{iB}) = \sum_{i} \left[\frac{n_{iA} n_{iB}}{n_i} \right] V_i^{-1} S_i,$$

in the sense that the variance of the linear combination of the corresponding differences in mean scores is minimized. This implies that the "optimal" linear combination of the S_i uses MVLE weights, given by

$$\omega_i \propto \left[\frac{n_{iA}n_{iB}}{N_i} \right] V_i^{-1}, \qquad (8.12)$$

which yield an optimal linear combination of the S_i, under a population model.

Weights may also be generated under other criteria. Rather than assuming a common mean difference in the scores as the alternative hypothesis of interest, one might consider a less restrictive alternative of stochastic ordering which specifies that the difference in mean scores is in the same direction for all strata. Wei and Lachin (1984) suggested a simple test of stochastic ordering that has been shown to be optimal (*cf.* Lachin (2000)). Their test is based on an unweighted combination of the mean differences over strata. Given the relationship in (8.11), this yields stochastic ordering test weights

$$\omega_i \propto \frac{n_i}{n_{iA}n_{iB}}. \qquad (8.13)$$

In general the degree of evidence against the null hypothesis, as represented by the Z-test statistic, will increase with the sample size. Thus rather than weight the test statistics (the numerators S_i), an alternate approach would be to weight the stratum-specific Z statistics, *i.e.*, the W_i, where $W_i = S_i/V_i^{1/2}$, such as

$$W_W = \frac{\sum_i \omega_i W_i}{\sqrt{\sum_i \omega_i^2}}$$

where W_W is distributed as $N(0,1)$ under H_0. For simple rank scores then one might simply weight by $\omega_i = n_i$. In the analysis of survival times using logrank scores, one might weight by $\omega_i = D_i$ where D_i is the number of deaths (events) in the ith stratum, which provides the degree of information in such data.

In each case, under the null hypothesis, the aggregate stratified-adjusted linear rank test in (8.8) will have type I error level α when the rejection region is based on the one or two-sided $1 - \alpha$ critical values, regardless of the set of weights employed. However, the result will depend on the chosen weights when the alternative hypothesis is true. In this case, the set of weights that yields the largest test value is that which is based on the alternative hypothesis best reflected in the data. Since the nature of the alternative hypothesis is unknown, then the optimal weights are also unknown. It would be cheating to conduct preliminary tests to select the weights that maximize the test. Lachin (2000, Section 4.9) presents a discussion of these issues.

The discussion in Chapter 7 regarding the relative merits of conditional and unconditional inference is also relevant in the context of stratified analyses. For a conditional test, one conditions on the ancillary statistic $N_A(n)$. In the stratified test, one conditions on the number of treatments assigned to A within each stratum. Flyer (1988) compares the power of a stratified test under a conditional randomization model and an unconditional randomization model and finds significant gains in power for the unconditional test. Nevertheless, many find the conditional test more appealing intuitively since the observed results are evaluated with reference to other random sequences with the same sample sizes as those obtained under the randomization employed in the study.

8.2.3 Small strata

In most trials the randomization is stratified by clinical center, and often there is a large disparity in the size of the strata, with some clinics recruiting a small number of subjects. In many instances, these small strata are pooled to form one strata of size comparable to that of other clinics. However, there is no randomization basis for doing so, and as we show in the following section, a test ignoring the strata employed in the randomization, if anything, is likely to be anti-conservative. Further, small strata still contribute to the test of the group difference, provided that at least one subject in each stratum is assigned to each group. In the case that no subjects are assigned to a group within a stratum, then that stratum cannot contribute to a comparison of treatments under a randomization model and should be discarded.

8.3 STRATIFIED VERSUS UNSTRATIFIED TESTS WITH STRATIFIED RANDOMIZATION

If one adopts a randomization basis for inference, then one should conduct the proper large-sample randomization-based test according to the particular randomization procedure used. For example, for a stratified randomization by clinic, the proper randomization analysis is also stratified by clinic. Such a stratified analysis will likely yield a test statistic which is different from that obtained by a simple aggregate analysis ignoring strata under either a population or randomization model.

The question now arises as to the difference, if any, between a simple unstratified analysis when the randomization was in fact blocked or stratified. This issue most commonly arises in the analysis of a study in which a permuted block design was employed, but where the analysis conducted ignores the blocking in computing the test statistic.

Assume that a permuted block design is employed with M blocks of assignments with a random allocation rule using block sizes $m = n/M$. If the analysis conducted ignores the blocking, the unblocked linear rank test statistic is

$$W_U = \frac{\sum_{i=1}^{M} \sum_{j=1}^{m} (a_{ij} - \bar{a}) T_{j(i)}}{\left[\frac{mM}{4(mM-1)} \sum_{i=1}^{M} \sum_{j=1}^{m} (a_{ij} - \bar{a})^2\right]^{1/2}}, \tag{8.14}$$

where a_{ij} is the rank score of the jth subject in the ith stratum with respect to the scores of all n subjects, and $\bar{a}$ is the total mean of all n scores. Thus $a_{ij} \neq a_{j(i)}$ in (8.9). If the random allocation design had been used in a single block with total $n = mM$, then this statistic would asymptotically be distributed as a standard normal

Table 8.1   Sum of squares and degrees of freedom (df) for a blocked analysis of variance with response variable $y_{j(i)}$ for the jth patient in stratum i.

Effect	Sum of Squares	df
Between blocks	$\sum_{i=1}^{M} m(\bar{y}_i - \bar{y})^2$	$M - 1$
Within blocks	$\sum_{i=1}^{M} \sum_{j=1}^{m} (y_{j(i)} - \bar{y}_i)^2$	$mM - M$
Total	$\sum_{i=1}^{M} \sum_{j=1}^{m} (y_{j(i)} - \bar{y})^2$	$mM - 1$

as shown in (8.7) above. However, under the permuted block design, the distribution of (8.14) is not necessarily the standard normal.

Squaring (8.9) and (8.14), their ratio yields

$$\frac{W_U^2}{W_B^2} = \frac{\left[\sum_{i=1}^{M} \sum_{j=1}^{m} (a_{ij} - \bar{a})T_{j(i)}\right]^2 \left[\frac{m}{4(m-1)} \sum_{i=1}^{M} \sum_{j=1}^{m} (a_{j(i)} - \bar{a}_i)^2\right]}{\left[\sum_{i=1}^{M} \sum_{j=1}^{m} (a_{j(i)} - \bar{a}_i)T_{j(i)}\right]^2 \left[\frac{mM}{4(mM-1)} \sum_{i=1}^{M} \sum_{j=1}^{m} (a_{ij} - \bar{a})^2\right]}.$$

When the blocked and unblocked scores are the same, $a_{ij} = a_{j(i)}$, then $\sum_{ij}(a_{ij} - \bar{a})T_{j(i)} = \sum_{ij}(a_{j(i)} - \bar{a}_i)T_{j(i)}$. This applies, for example, to binary scores in which case $W_U^2 = X_U^2$ in (8.3) and $W_S^2 = X_S^2$ (8.6). Then, noting that the sum of squares total equals the sum of squares blocks plus the sum of squares within blocks, as shown in Table 8.1, the relationship between the two statistics can be expressed as

$$1 - \frac{X_U^2}{X_B^2} = \frac{MSB - MSW}{MSB + \frac{M}{M-1}(m-1)MSW}, \tag{8.15}$$

where MSB is the block mean square of the a_{ij} and MSW is the within block mean square of the a_{ij} from an analysis of variance with just these two sources of variability. For large M, the above expression is the intrablock correlation coefficient, expressed as

$$R = \frac{MSB - MSW}{MSB + (m-1)MSW}. \tag{8.16}$$

Under both a population model and a randomization model, Matts and Lachin (1988) show that equivalent results apply to an analysis of variance of mean values when the proper blocked (stratified) chi-squared or F-test is compared to the unstratified test.

Thus, in the comparison of a blocked (stratified) versus unblocked Mantel-Haenszel analysis of 2×2 tables, or an analysis of variance, whether the unblocked test with X_U^2 is conservative, anti-conservative, or the same, when compared to the blocked test X_B^2, depends on whether the value of the intrablock correlation coefficient R is positive, negative, or zero, respectively. If R is zero, the two linear rank statistics are identical and thus the statistic which ignores the blocking (X_U^2) is also asymptotically distributed as χ^2 on 1 degree of freedom under the permuted block design.

If R is negative, then X_U^2 is stochastically larger than X_B^2, in which case the unblocked statistic X_U^2 will be anti-conservative. In this case the distribution of X_U^2 is stochastically larger than the χ^2 distribution on 1 degree of freedom and, as a result, both the type I error and power are inflated relative to the proper test X_B^2.

If R is positive, the X_U^2 will be conservative or stochastically smaller than X_B^2. In this case the distribution of X_U^2 is stochastically smaller than the χ^2 distribution on 1 degree of freedom and, as a result, both the type I error probability and power are deflated relative to the proper test X_B^2.

The range of possible values of the intrablock correlation is $-1/(m-1) \le R \le 1$. With a block size of $m = 2$, the lower bound for R is -1. With a block size of four, the lower bound is -0.33. As the block size increases, the lower bound for R approaches zero. Thus, as the block size increases, it is increasingly likely that an unblocked test will result in either a similar or a conservative test compared to the proper blocked permutation test.

An intrablock correlation of 0 would arise when patient responses are uniformly distributed over all subjects from 1 to n. A positive intrablock correlation would arise, for example, if the patients recruited early in a trial are healthier than those recruited later, or vice versa. A negative correlation would arise when the two treatment groups had time trends that were opposite, such as one having increasing values and the other decreasing values. With randomization this would be highly unlikely. Thus, a positive correlation, if any, is likely to occur in most trials, in which case the stratified blocked analysis should be performed to obtain a test of the proper size.

8.4   EFFICIENCY OF STRATIFIED RANDOMIZATION IN A STRATIFIED ANALYSIS

In Section 8.5 to follow, we will consider post-stratification in the analysis on the basis of a covariate that was not employed in the stratification or blocking of the initial randomization, and the subsequent analysis. One question that often arises in the planning of a clinical trial is whether a stratified randomization will increase the precision of a stratified analysis. The relative efficiency of a stratified test with stratified randomization versus without stratified randomization was assessed by Grizzle (1982) using a homogeneous population model, and by Matts and McHugh (1978) using a randomization model. Essentially identical results were obtained.

Following Grizzle's treatment, assume that the subjects arise from a homogeneous population over time within each stratum. For example, if patient gender is the covariate of interest, whether in a study with randomization stratified by gender or one unstratified, we assume that the likelihood that a male will enter the study is a constant over time. This assumption is equivalent to assuming that the covariate values arc independently and identically distributed within each stratum.

Grizzle assessed this issue as follows. Consider the simplest case of two treatments ($k = A$ or B) and two strata ($i = 1, 2$) in a simple additive linear model

$$Y_{ijk} = \alpha + \beta_i + \mu_k + \epsilon_{ijk}, \tag{8.17}$$

for $j = 1, ..., n_{ik}$ subjects on treatment k in stratum i. As usual, the errors are assumed to be independent and identically distributed with $E(\epsilon_{ijk}) = 0$ and $\mathrm{Var}(\epsilon_{ijk}) = \sigma^2$. The effect of the kth treatment is μ_k, and that of the ith stratum is β_i, with $\beta_1 + \beta_2 = 0$. The treatment effect is $\theta = \mu_1 - \mu_2$. It is important to note that there is no treatment-stratum interaction. Thus, treatment effects within strata are assumed to be homogeneous, thus maximizing the gains in power from a stratified analysis.

In the case of a stratified randomization, it is assumed that the sample sizes allocated to each treatment are always equal, either in total or within strata. That is, we assume that the stratified randomization was 100 percent effective in eliminating covariate imbalances on the stratifying covariate(s). This is guaranteed if a random allocation rule or permuted-block randomization is employed (with all blocks filled), and is the expectation with the other randomization procedures (complete or urn randomization).

Now consider the case of an unstratified randomization. Let $q_k = n_{1k}/n_k, k = A, B$, be the proportion of subjects randomized to the kth group who are also members of the first stratum ($i = 1$), and $1 - q_k$ be the proportion of those in the kth group who are members of the second stratum ($i = 2$). A covariate imbalance occurs when $q_A \neq q_B$. Conversely, with 100 percent-effective stratified randomization, it is assumed that the covariate stratum fractions are fixed and equal, $q_A = q_B$, so that there is no covariate imblance.

Now consider the efficiency of an estimator of θ. Denote the variance of the estimator with stratified randomization (r) and stratified analysis (a), such that $q_A = q_B$, as $\sigma^2_{\theta(r,a)}$. Likewise, denote the variance of the estimator with unstratified randomization but with a subgroup analysis as $\sigma^2_{\theta(a)}$. When $n_A = n_B$, the relative efficiency of the estimators is then

$$R.E. = \frac{\sigma^2_{\theta(r,a)}}{\sigma^2_{\theta(a)}}.$$

Using the least squares estimator from (8.17), Grizzle (1982) shows that

$$\sigma^2_{\theta(a)} = \frac{4\sigma^2}{n} \left[\frac{q_A + q_B}{q_A(1 - q_A) + q_B(1 - q_B)} \right] \left[1 - \frac{q_A + q_B}{2} \right]. \tag{8.18}$$

Then with stratified randomization, taking $q_A = q_B$ in (8.18), we obtain

$$\sigma^2_{\theta(r,a)} = \frac{4\sigma^2}{n}.$$

The relative efficiency is then given by

$$R.E. = \left\{ \frac{q_A + q_B}{q_A(1 - q_A) + q_B(1 - q_B)} \left[1 - \frac{q_A + q_B}{2} \right] \right\}^{-1}. \tag{8.19}$$

Table 8.2 *Relative efficiency of estimators for stratified randomization and stratified analysis versus stratified analysis only, for various values of q_A and q_B.*

q_A	q_B	R.E.
0.3	0.1	0.938
0.5	0.1	0.857
0.5	0.3	0.821
0.7	0.1	0.625
0.7	0.3	0.840
0.7	0.5	0.958
0.9	0.1	0.360
0.9	0.3	0.625
0.9	0.5	0.810
0.9	0.7	0.938

Note that $R.E. \leq 1$ and $R.E. = 1$ when $q_A = q_B$. This relative efficiency of the estimators is also proportional to the relative power of a statistical test of $H_0 : \theta = 0$ using a post-hoc stratified analysis versus a stratified randomization and a stratified analysis. (Grizzle (1982) also gives the relative error for the case where $n_A \neq n_B$.) From equation (8.19), Table 8.2 can be computed which gives the relative efficiencies for various values of q_A and q_B ranging from 0.10 to 0.90. Lachin (2000, Section 3.5.4) presents an equivalent model for the analysis of binary data.

Now, suppose that q_A and q_B are binomial random variables with $E(q_A) = E(q_B) = \gamma$. For a given value γ, we can use the normal approximation to compute the probability that their absolute difference exceeds some value r. This is given by

$$\Pr\left(|q_A - q_B| > r\right) = 2\left\{1 - \Phi\left(\frac{r\sqrt{n}}{2\sqrt{\gamma(1-\gamma)}}\right)\right\}.$$

For various values γ ranging from 0.1 to 0.9, and for various sample sizes n, Table 8.3 gives the limits of imbalance which would occur with probability 0.05 and 0.01. These imbalances can then be used with Table 8.2 to assess the loss of efficiency due to non-stratification.

For example, for $n = 25$ and $\gamma = 0.5$, an imbalance of 0.7 and 0.3 could occur with $p \cong 0.05$, which would result in an efficiency of 0.84 with unstratified randomization. However, for $n = 100$, there is probability < 0.01 of covariate imbalances which would result in a relative efficiency of 0.9 or less.

The above results apply regardless of the method of randomization or treatment assignment employed because a homogeneous population model is assumed. A randomization-based analysis of this same issue was explored by Matts and McHugh (1978) assuming that a random allocation rule was employed with and without stratification. Again, note that the random allocation rule guarantees equal sample

Table 8.3   *Limits of imbalance occurring with probability* 0.05 *and* 0.01, *for various values of n and* γ.

n	γ	0.05	0.01
25	0.1	0.235	0.309
25	0.3	0.359	0.472
25	0.5	0.392	0.515
50	0.1	0.166	0.219
50	0.3	0.254	0.334
50	0.5	0.277	0.364
100	0.1	0.116	0.154
100	0.3	0.180	0.236
100	0.5	0.196	0.258
200	0.1	0.083	0.109
200	0.3	0.127	0.167
200	0.5	0.139	0.182

sizes within each group in total and within each stratum. Matts and McHugh also use a simple linear model like (8.17), but allow for more than two treatment groups and an arbitrary number of strata.

For the case of only two equal sized groups, they show that the relative efficiency for a study of size n, with s strata of equal size $n_i = n/s$, is obtained as

$$R.E. = \frac{n\left[1 - \left(1 - \frac{1}{s}\right)^n\right]}{n + 2s - 2}. \tag{8.20}$$

Clearly, as n increases relative to s, $R.E. \to 1$. Solving for n in (8.20) and ignoring the asymptotically negligible term $\{(s-1)/s\}^n$, we obtain

$$n \cong \frac{2R.E.(s-1)}{1 - R.E.}.$$

For example, for $s = 10$ a stratified analysis with unstratified randomization will yield 90 percent of the efficiency of a stratified randomization with a stratified analysis for a sample size of $n \cong 162$ or greater. For $s = 10$, 95 percent efficiency is provided by $n \cong 342$ or greater.

Therefore, under two entirely different approaches, it has been shown that a stratified randomization will have non-negligible effects on the efficiency of a stratified analysis with small sample sizes, but that as the sample size increases, there are miniscule gains in efficiency from a stratified randomization relative to a simple post-hoc stratified analysis. Unfortunately, as discussed in Chapter 4, even though stratification has greatest merit in small trials, it is usually not feasible to stratify on more than one or two factors due to the small within-stratum cell sizes.

8.5   POST-HOC STRATIFIED AND SUBGROUP ANALYSES

In addition to the overall comparison of treatments A and B, it may also be desired to compare treatments separately among those patients who are members of a subgroup defined post-hoc on the basis of a covariate, usually a baseline (pre-randomization) characteristic not used as a basis for stratification in the randomization. Such post-stratified analyses are used to obtain a stratified-adjusted assessment of the overall treatment effect. For example, the treatments might be compared separately among men and separately among women in a study where the randomization was not stratified by gender. A covariate-adjusted test of treatment effect might then be obtained by combining the separate tests for men and women.

While such stratified analyses may in fact be specified *a priori* (in fact, such specification is preferred), such analyses are post-hoc with respect to the generation of the randomization. Any such analysis specified after examination of the data could be criticized unless the basis for the analysis were specified *a priori*. For example, the protocol might specify that a post-hoc stratified analysis would be conducted to adjust for any covariates on which the groups differed significantly by chance.

In order to perform a valid analysis among the subsets of patients within such subgroups, it is sufficient to assume that the randomly assigned treatment indicator variable values $\{T_j\}$ are statistically independent of the covariate values $\{X_j\}$ among the n patients randomized. We refer to this as the *covariate independence assumption*, which specifically assumes that $E(T_j|\mathcal{F}_{j-1}, X_j) = E(T_j|\mathcal{F}_{j-1})$ where $\mathcal{F}_{j-1}$ is the history of prior allocations, as would apply to a restricted randomization procedure. Clearly this assumption is satisfied for any baseline covariate when there is no potential for selection bias. It does not apply to covariate-adaptive randomization procedures.

For complete randomization, since the probability of treatment assignment to A is $E(T_j) = 1/2$ independently for all patients, the permutation test can be performed within any subgroup as though a separate randomization had been performed within that subgroup. Further, the test statistics within each of multiple strata will be statistically independent, and thus can be combined using (8.8), exactly as though stratified randomization had been performed. For any other randomization procedure, since the probabilities of assignment are not independent and identically distributed, the validity of such analyses rests on the covariate independence assumption. For example, this assumption could be violated if the randomization is open to selection bias.

There are two ways that a post-stratified analysis could be performed. For simplicity assume that the original randomization was unstratified (later we also consider the case of a pre-stratified randomization). In a *stratum-centered scores analysis* the scores a_j are computed for all n subjects according to the original randomization; *i.e.* unstratified in this case. Then these scores are used to compare groups within strata by computing a sum of deviations from the within-stratum mean and using the sum of squares about that mean in the variance. For example if there are four subjects in a stratum with unstratified rank scores $(2, 5, 8, 13)$, then these values with mean 7 could be used to compute a rank statistic within that stratum. If we use an indicator variable $\nu_{\ell j} = 1$ or 0 to denote patient j in the ℓth stratum, $\ell = 1, ..., L$, then the rank statistic within that stratum would be of the form

$$S_\ell = \sum_{j=1}^{n} \nu_{\ell j}(a_j - \bar{a}_\ell)T_j, \tag{8.21}$$

where the $\{a_j\}$ are the unstratified rank scores (for notational simplicity, we drop the dependence on n) and $\bar{a}_\ell$ is the mean score within stratum ℓ. This construction would mimic a post-stratified analysis of mean values in an analysis of variance wherein the response values are unchanged, but a within stratum or block mean correction is employed in computing the sums of squares.

However, if the randomization had been pre-stratified by a factor, then the appropriate approach would be to compute rank scores within each stratum as in (8.8). This *stratum-specific scores analysis* can also be applied to strata defined post-hoc whereby the rank scores are computed separately within each stratum. In this case the post-stratified rank scores for the above example would be $(1, 2, 3, 4)$ with mean 2.5.

Clearly either approach is valid and the relative merits of one versus the other have not been explored, to our knowledge. Lachin (1988) and Matts and Lachin (1988) employ a stratum-specific analysis following compete randomization (conditionally) and a random allocation rule design, as well as permuted blocks. Wei and Lachin (1988) employ a stratum-centered scores analysis following Wei's urn design.

8.5.1 Complete randomization

Without any loss of generality, assume that unstratified complete randomization was employed to assign treatments A versus B to n patients in a clinical trial. A post-hoc stratified analysis is then performed within each of L multiple mutually exclusive subgroups (strata) defined post-hoc on the basis of a covariate. Here we use L to denote the number of post-hoc strata in distinction to the number of pre-hoc strata denoted previously by I. The following is based on developments in Lachin (1988).

For simplicity, consider the case of two strata with indicator variables ν_{1j} and ν_{2j} to indicate membership in stratum 1 ($\nu_{1j} = 1, \nu_{2j} = 0$) or in stratum 2 ($\nu_{1j} = 0, \nu_{2j} = 1$) for the jth patient, $j = 1, \ldots, n$. Note that $\nu_{1j} + \nu_{2j} = 1$ and that $\nu_{1j}\nu_{2j} = 0$ for all j. For the jth patient, the stratum-specific rank score is defined as some function of the responses among members of that patient's subgroup, most generally as

$$c_j = \nu_{1j}f(\nu_{11}, Y_1, \ldots, \nu_{1n}, Y_n) + \nu_{2j}f(\nu_{21}, Y_1, \ldots, \nu_{2n}, Y_n). \tag{8.22}$$

These scores are equivalent to those that would have been computed had the randomization been stratified by this factor as described in Section 8.2.2. However a different notation is employed to distinguish post-stratified scores from pre-stratified scores $a_{j(i)}$.

In keeping with the principles of the randomization model, we assume that subgroup membership is deterministic, as are the scores. The only randomness is in the treatment assignments $\{T_j\}$. For the ℓth subgroup ($\ell = 1, 2$), the total sample size is $n_\ell = \sum_{j=1}^{n} \nu_{\ell j}$, of whom $N_{\ell A} = \sum_{j=1}^{n} \nu_{\ell j} T_j$ are assigned to treatment A, and $N_{\ell B} = n_\ell - N_{\ell A}$ are assigned to treatment B. The above generalizes naturally for the case of $L > 2$ strata. Then the linear rank statistic within the ℓth subgroup can be written as

$$W_\ell = \frac{S_\ell}{\sqrt{\mathrm{Var}(S_\ell)}}, \tag{8.23}$$

where

$$S_\ell = \sum_{j=1}^{n} \nu_{\ell j}(c_j - \bar{c}_\ell) T_j$$

and where

$$\bar{c}_\ell = \frac{\sum_{j=1}^{n} \nu_{\ell j} c_j}{n_\ell}$$

for $\ell = 1, 2$. It is straightforward to verify that $E(S_\ell) = 0$ since $\sum_{j=1}^{n} \nu_{\ell j}(c_j - \bar{c}_\ell) = 0$, and that

$$V_\ell \equiv \mathrm{Var}(S_\ell) = \frac{1}{4} \sum_{j=1}^{n} \nu_{\ell j}(c_j - \bar{c}_\ell)^2. \tag{8.24}$$

Suppose that n_ℓ grows large as n grows large. Then from Section 14.2.1, the following condition will ensure asymptotic normality of W_ℓ:

$$\lim_{n \to \infty} \frac{\max_{1 \le j \le n} \nu_{\ell j}(c_j - \bar{c}_\ell)^2}{\sum_{j=1}^{n} \nu_{\ell j}(c_j - \bar{c}_\ell)^2} = 0. \tag{8.25}$$

Now let us examine the conditional permutation test, conditional on $N_{\ell A}$. Then $E(\nu_{\ell j} T_j | N_{\ell A}) = N_{\ell A}/n_\ell$ and

$$
\begin{aligned}
V_\ell &= \sum_{j=1}^{n} \nu_{\ell j}(c_j - \bar{c}_\ell)^2 \mathrm{Var}(\nu_{\ell j} T_j | N_{\ell A}) \\
&+ \sum_{j=1}^{n} \sum_{\substack{i=1 \\ i \ne j}}^{n} \nu_{\ell j} \nu_{\ell i}(c_j - \bar{c}_\ell)(c_i - \bar{c}_\ell) \mathrm{cov}(\nu_{\ell j} T_j, \nu_{\ell i} T_i | N_{\ell A}).
\end{aligned} \tag{8.26}
$$

Since $\mathrm{Var}(\nu_{\ell j} T_j | N_{\ell A}) = N_{\ell A} N_{\ell B}/n_\ell^2$, and for $i \ne j$,

$$\mathrm{cov}(\nu_{\ell j} T_j, \nu_{\ell i} T_i | N_{\ell A}) = -N_{\ell A} N_{\ell B}/n_\ell^2(n_\ell - 1),$$

we have from (8.26) that

$$V_\ell = \frac{N_{\ell A} N_{\ell B}}{n_\ell(n_\ell - 1)} \sum_{j=1}^{n} \nu_{\ell j}(c_j - \bar{c}_\ell)^2. \tag{8.27}$$

These expressions then provide for a post-hoc randomization-based test of $H_{0\ell}$ within the ℓth subgroup using a subgroup specific statistic. By the results of Section 14.2.2, if $n_\ell \to \infty$ as $n \to \infty$, if (8.25) holds, and if

$$\lim_{n \to \infty} \frac{\sum_{j=1}^{n} \nu_{\ell j}(c_j - \bar{c}_\ell)^2}{n} \to \gamma, \tag{8.28}$$

for a constant γ, then

$$W_\ell = \frac{S_\ell}{\sqrt{\gamma N_{\ell A} N_{\ell B}/n_\ell}}$$

is asymptotically standard normal.

We also wish to conduct an aggregate post-hoc stratified adjusted test of significance. To do so requires that we first explore the covariance matrix of the set of subgroup-specific statistics. Without loss of generality consider the case of $L = 2$ subgroups or post-hoc defined strata. We now show that the covariance between the rank statistics S_1 and S_2 is zero, and therefore, that the statistics are independent. This is clear for the unconditional test, since S_1 and S_2 depend on only on the unconditional distribution of the treatment assignments, and each treatment assignment can appear only in S_1 or S_2.

Over the conditional reference set, we can compute

$$\text{cov}(S_1, S_2 | N_{1A}, N_{2A})$$

$$= \sum_{j=1}^{n} \nu_{1j} \nu_{2j} (c_j - \bar{c}_1)(c_j - \bar{c}_2) \text{Var}(T_j | N_{1A}, N_{2A})$$

$$+ \sum_{j=1}^{n} \sum_{\substack{i=1 \\ i \neq j}}^{n} \nu_{1j} \nu_{2i} (c_j - \bar{c}_1)(c_i - \bar{c}_2) \text{cov}(T_j, T_i | N_{1A}, N_{2A}).$$

The first term is zero since $\nu_{1j} \nu_{2j} = 0$. The second term is zero since

$$\text{cov}(\nu_{1j} T_j, \nu_{2i} T_j | N_{1A}, N_{2A}) = 0,$$

as treatments are independent in different strata. The latter will *not* be the case for some restricted randomization schemes, as we will discuss later.

Consequently, it is possible to perform a post-stratified covariate-adjusted combined test using a slight modification of (8.8) with ℓ used to index strata rather than i. For the unconditional test, the aggregate test

$$W_{PS} = \frac{\sum_{\ell=1}^{L} \omega_\ell S_\ell}{\left[\sum_{\ell=1}^{L} \omega_\ell^2 V_\ell\right]^{\frac{1}{2}}}, \tag{8.29}$$

where V_ℓ is defined by (8.24) and with weights ω_ℓ, is asymptotically normal, provided (8.25) holds. For the conditional test,

$$W_{PS} = \frac{\sum_{\ell=1}^{L} \omega_\ell S_\ell}{\left[\sum_{\ell=1}^{L} \omega_\ell^2 \gamma N_{\ell A} N_{\ell B}/n_\ell\right]^{\frac{1}{2}}} \tag{8.30}$$

will be asymptotically normal if (8.25) and (8.28) hold. Here the weights ω_ℓ are specific to each post-hoc specified subgroup and may be chosen in a similar manner as for a stratified analysis following stratified randomization.

8.5.2 Random allocation rule

For the random allocation rule, the probabilities of assignment within a given sequence are a function of the prior assignments. However, because the sample sizes are fixed *a priori*, usually as $n/2$ to each group, then each permutation within the reference set is equiprobable. Likewise, if it is assumed that subgroup membership has equal probabilities in each treatment group, then conditional on the sample sizes ($N_{\ell A}$ and $N_{\ell B}$) within the ℓth subgroup, each permutation within the reduced reference set for that subgroup is also equiprobable. That is, all possible permutations of n_ℓ out of n patients are equally likely, as are all possible permutations of $N_{\ell A}$ out of n_ℓ assignments to treatment A. Therefore, the randomization variance within the ℓth subgroup is also given by (8.29). Further, since the conditional complete randomization variance-covariance structure also applies to assignments generated from a random allocation rule, then the statistics S_1 and S_2 are independent. Thus all the above developments also apply to a post-hoc stratified analysis following assignments using the random allocation rule (*cf.* Lachin, 1988).

Post-hoc stratified analysis following a truncated binomial design has not been explored.

8.5.3 Permuted block randomization with a random allocation rule

Under the covariate independence assumption, a permutation test can likewise be performed with a random allocation rule permuted-block design using only the responses from patients within each block who are members of the designated subgroup. For binary observations, Matts and Lachin (1988) show that a block-stratified Mantel-Haenszel subgroup analysis will provide a test equivalent to the permutation test. For quantitative observations, the blocked analysis of variance using responses only from members of the subgroup will provide an F-test which is asymptotically equivalent to the permutation test.

For the family of linear rank tests, the proper permutation test within a subgroup under the covariate-treatment independence assumption is a generalization of (8.29). Conditional on the pattern of subgroup indicators within each block $\nu_{\ell ij}$, for stratum $\ell = 1, ..., L$, block $i = 1, ..., M$, each of size m, and patient $j = 1, ..., n$. The stratum-specific scores $\{c_{ij}\}$ are defined as in (8.22) as a function of the responses Y_{ij} within each block, and the stratum means are given by $\bar{c}_{\ell i}$. Then the variance of the rank statistic for the ℓth subgroup within a block is the same as that for the random allocation rule. Therefore, conditional on the pattern of subgroup indicators within the entire trial, the linear rank test within the ℓth subgroup becomes

$$
\begin{aligned}
W_\ell &= \frac{S_\ell}{\sqrt{V_\ell}} \\
&= \frac{\sum_{i=1}^{M} \omega_{\ell i} \sum_{j=1}^{m} \nu_{\ell i j}(c_{ij} - \bar{c}_{\ell i})T_{ij}}{\left[\sum_{i=1}^{M} \omega_{\ell i}^2 \frac{N_{\ell i A} N_{\ell i B}}{N_{\ell i}(N_{\ell i}-1)} \sum_{j=1}^{m} \nu_{\ell i j}(c_{ij} - \bar{c}_{\ell i})^2 \right]^{1/2}},
\end{aligned}
\tag{8.31}
$$

where $N_{\ell i A} = \sum_{j=1}^{n} \nu_{\ell i j} T_{j(i)}$ and $n_{\ell i} = \sum_{j=1}^{n} \nu_{\ell i j}$ are the numbers of subgroup members in group A and in total in block i who belong to subgroup ℓ, respectively, T_{ij} is the treatment assignment of the jth patient in the ith block, and $\omega_{\ell i}$ is a weight associated with the block. Note that the weights may differ for each block and each subgroup depending on the sample sizes and the nature of the weights and scores for each subgroup within each block. The asymptotic theory follows from Sections 8.5.1 and 8.5.2 for large M.

In the event that multiple mutually exclusive subgroups are defined on the basis of a covariate, then under the covariate independence assumption, it then follows that the rank statistics for each subgroup are statistically independent. Thus, a combined, covariate-adjusted test can be performed as in (8.29).

Post-hoc stratified analysis following permuted block randomization using a truncated binomial design has not been explored.

8.5.4 Wei's urn design

While the development of post-hoc stratified analyses for complete randomization and the random allocation rule (with extension to permuted block designs) has been straightforward, the literature has largely not addressed more complicated restricted randomization procedures, such as the truncated binomial design and Efron's biased coin design. Davis (1986) develops the theory for a post-hoc stratified *unconditional* test following Wei's urn design. The form of the *conditional* test statistic and its distribution theory appears to be an open problem. In this section, we address the unconditional test for Wei's urn design. The main additional complication with these more complicated restricted randomization procedures is that stratum-specific linear rank tests will not be independent.

While Davis (1986) develops the test using the more complicated expression for the asymptotic variance described in Section 14.6, it is clear from simulations presented there that the form of the variance from complete randomization can be substituted without any serious departures from normality. Hence the form of the stratum-specific test in (8.23) should be appropriate for the unconditional stratum-specific test.

The distribution of the post-hoc stratified tests follows from developments in Davis (1986). Let $\boldsymbol{\nu}_j = (\nu_{1j}, ..., \nu_{Lj})'$ and let $c_j - \bar{c}_\ell$ be the centered scores, which may be stratum-specific scores or stratum-centered scores. Further, let $\boldsymbol{b}_j = \boldsymbol{\nu}_j(c_j - \bar{c}_\ell)$.

Then we define an $L \times L$ matrix

$$\Lambda = \frac{1}{4} \sum_{j=1}^{n} b_j b_j'.$$

Let $S = (S_1, ..., S_L)$ be the vector of stratum-specific linear rank statistics, where

$$S_\ell = \sum_{j=1}^{n} \nu_{\ell j}(c_j - \bar{c}_\ell)T_j, \ell = 1, ..., L.$$

Then under the condition

$$\lim_{n \to \infty} \frac{\max_{1 \leq j \leq n} (c_j - \bar{c}_\ell)^2}{\min_{1 \leq \ell \leq L} \sum_{j=1}^{n} \nu_{\ell j}(c_j - \bar{c}_\ell)^2},$$

we have that $\Lambda^{-1/2} S$ is asymptotically multivariate normal with mean 0 and variance-covariance matrix I. This provides a separate test of $H_{0\ell}$ within the subgroup of the form $W_\ell = S_\ell / \sqrt{V_\ell}$, where $V_\ell = \Lambda_{\ell\ell}$.

Furthermore, if there is no obvious interaction between treatment and strata, then the $\{S_\ell\}$ can be combined in a linear fashion using a vector of weights $\omega = (\omega_1 ... \omega_L)'$ in

$$W_{PS} = \frac{\omega' S}{(\omega' \Lambda \omega)^{1/2}}, \tag{8.32}$$

in order to make an overall inference about the treatment difference.

As before, the choice of the weights would depend on the alternative of interest in some respect. In this case, since the S_ℓ are correlated, then the weights are different from those described in Section 8.2.2. If it is assumed that there is a common difference in the mean scores over strata, then the generalized least squares weighted combination uses a weight vector

$$\omega = C \Lambda^{-1} C,$$

where C is a diagonal matrix with elements

$$c_{\ell\ell} = \frac{N_{\ell A} N_{\ell B}}{n_\ell}.$$

Conversely, a test directed to the alternative of stochastic ordering would use weights as in (8.13) substituting ℓ for i.

8.5.5   Pre- and post-stratified analyses

The preceding methods can be generalized in an obvious manner to a post-stratified analysis on one factor (say F) with L levels in a trial that employed a randomization stratified by another factor (say G) with I levels. In this case, the post-stratified analysis on F is conducted separately within each level of G, to generate a rank statistic $S_{\ell(i)}$ for the ℓth level of F within the ith level of G, with variance $V_{\ell(i)}$ using generalizations of the above. Then the results are pooled over levels of F and/or G as appropriate.

For complete randomization or a random allocation rule, including permuted blocks, a test of the treatment effect within the ℓth level of F is obtained as

$$W_\ell = \frac{S_\ell}{V_\ell^{1/2}} = \frac{\sum_{i=1}^{I} \omega_{\ell(i)} S_{\ell(i)}}{\left[\sum_{i=1}^{I} \omega_{\ell(i)}^2 V_{\ell(i)}\right]^{1/2}}, \tag{8.33}$$

which is a generalization of (8.29) or (8.30) with weights $\omega_{\ell(i)}$ for the ℓth level of F within the ith level of G. Then to obtain an overall stratified-adjusted test, adjusting for both F and G, the S_ℓ and V_ℓ in (8.33) are employed in (8.29). The expression in (8.31) for a permuted block randomization is a special case of this general approach.

For Wei's urn design, within the ith level of the pre-randomization stratification factor G, the post-hoc stratified analysis yields a vector of statistics $\mathbf{S}_i = (S_{1i} \ldots S_{Li})'$ with $L \times L$ covariance matrix $\mathbf{\Lambda}_i$, for $i = 1, \ldots, I$. Then these vectors can be combined over strata using a symmetric weight matrix $\mathbf{\Omega}_i$ such that

$$\mathbf{S} = \sum_{i=1}^{I} \mathbf{\Omega}_i \mathbf{S}_i \tag{8.34}$$

with variance-covariance matrix

$$\mathbf{\Lambda} = \sum_{i=1}^{I} \mathbf{\Omega}_i \mathbf{\Lambda}_i \mathbf{\Omega}_i. \tag{8.35}$$

These can then be employed in (8.32) with an appropriate weight vector to provide a stratified-adjusted test of H_0.

Using the developments in Section 8.2.2, an appropriate weight matrix for the ith stratum is

$$\mathbf{\Omega}_i = \mathbf{C}_i \mathbf{\Lambda}_i^{-1} \mathbf{C}_i, \tag{8.36}$$

where $\mathbf{C}_i$ is a diagonal matrix with elements

$$c_{i\ell} = \frac{N_{i\ell A} N_{i\ell B}}{n_{i\ell}} \tag{8.37}$$

for $i = 1, \ldots, I$. The $N_{i\ell A}, N_{i\ell B}$ and $n_{i\ell}$ are the sample sizes within each group and total within the ℓth level of F and the ith level of G.

The above construction provides a separate test within each joint stratum defined by levels of F and G jointly, aggregate tests within pre-randomization strata, and then aggregate tests over all strata combined. If the latter test is the only test of interest, then an alternate approach would be to construct one large vector $\mathbf{S}$ consisting of the IL stratum specific statistics. These could then be combined in one step to provide a jointly stratified test. This approach is left to a problem.

8.6 ANALYSES WITH MISSING DATA

A related concern is the validity of an analysis when responses $\{Y_j\}$ for some patients are missing. Often patients with missing data are simply excluded and the analysis is based only on the subset of patients with complete data. Under a population model, such an analysis can be justified when it can be assumed that the missing data are missing-completely-at-random (MCAR), or that missingness is statistically independent of the observable response (*cf.* Little and Rubin, 1987). Under a randomization model an analogous assumption is the covariate independence assumption, as defined in Section 8.5, where the post-hoc covariate is defined as having two categories: missing or observed.

Let Y_j be the potentially observable response, and let ν_j be an indicator variable to denote whether the response of the jth patient is observed ($\nu_j = 1$) or missing ($\nu_j = 0$). Since we will only conduct an analysis within the subgroup with observed data, only one subgroup indicator variable is employed.

Under a population model, the treatment assignments T_j are deterministic and under the MCAR assumption, the responses $\{Y_j\}$ are statistically independent of the $\{\nu_j\}$. Under a randomization model, however, the responses Y_j are fixed and it is assumed that the treatment assignments $\{T_j\}$ are statistically independent of the $\{\nu_j\}$ (see Lachin, 1988). This implies that the expected probability of treatment assignment is the same for those patients with observed data and those with missing data.

Unfortunately, under either a population model with the MCAR assumption, or a randomization model with the independence assumption, there is no possible direct test for the required assumptions. These assumptions can only be assessed indirectly, such as by examining the observed characteristics of those patients with missing data versus those with observed data. For this reason, it is important that the incidence of missing data be kept to a minimum.

When no observations are missing, the rank score a_j can be written as some function $a_j = f(Y_1, \ldots, Y_n)$. In the presence of missing data, however, analogous to (8.22) the rank score is defined as $c_j = f(\nu_1, Y_1, \ldots, \nu_n, Y_n)$; i.e., c_j is defined as a function of the complete observations only and is undefined if Y_j is missing. The numerator of the linear rank test then becomes $S = \sum_{j=1}^{n} \nu_j (c_j - \bar{c}) T_j$ where $\bar{c} = (\sum_{j=1}^{n} \nu_j c_j)/(\sum_{j=1}^{n} \nu_j)$ and the denominator involves $V = Var(S)$.

With complete randomization, a randomization using a random allocation rule, or Wei's urn design, the statistical considerations are the same as for a subgroup analysis since the subset of patients with observed data is exactly such a post-hoc defined subgroup.

As described by Matts and Lachin (1988), with a permuted-block randomization any missing data can be viewed as a special case of an unfilled block. In this design it is statistically valid to exclude a block from the analysis due to operational deficiencies, such as missing data (unrelated to treatment effects), or incomplete recruitment (an unfilled block). This is termed a *complete-block randomization analysis.* Exclusion of such incomplete blocks will not affect the integrity of the remaining complete blocks, for which the resulting aggregate test statistic (with blocking) still has the usual randomization distribution. Thus, an unbiased and valid randomization analysis can be performed using the subset of complete blocks without the need to invoke any additional assumptions.

However, fewer patients would contribute to a complete-block randomization analysis than would contribute to a complete-data randomization analysis, thus potentially resulting in a loss in efficiency. Further, such a complete-block randomization analysis strictly should only be interpreted to apply to the collection of patients in the complete blocks. In order to apply the results to the original collection of n patients randomized, it is necessary to invoke the assumption of independent missing data. Under this assumption, a valid randomization analysis can also be performed using the subset of patients with complete data as a post-hoc subgroup. In this case the blocked subgroup analysis should be performed as described in (8.31) ignoring the subscript ℓ.

In some instances response measures are informatively missing for reasons that are obviously not random in a population model sense, or independently in a randomization model sense. A common example is where the outcome measure $\{Y_j\}$ is a measure of quality of life or disease severity but where some subjects die during the study. In this case, the fact that the subject died indicates the worst possible quality of life or disease severity. For such cases, Lachin (1999) describes a worst rank analysis in which subjects who die are assigned a common value more extreme than that of the observed values, such that all deaths share a tied worst rank. Lachin (1999) also describes a method for assigning rank scores in a time to event analysis such that the deaths have untied worst scores.

8.7   COVARIATE-ADJUSTED ANALYSES

A stratified-adjusted analysis is one approach to adjust for any bias introduced by an uncontrolled covariate, or to increase the efficiency by accounting for a highly influential covariate. This approach, however, is only applicable to qualitative covariates, or discretized quantitative covariates, and few in number. In many respects it is more natural to perform an adjustment using a regression model that allows for both qualitative and quantitative covariates simultaneously. While a regression model is usually developed under a population model, it is straightforward to apply a randomization analysis following the fit of a model. Conceptually the basic steps are to first fit a model to baseline covariates, other than treatment group. Then the residuals from the model can be viewed as a set of pre-ordained responses, regardless of which treatment is assigned. The residuals $\{e_j\}$ can then be employed in lieu of the responses $\{Y_j\}$ as the basis for computing a rank score.

Under a population model, the responses $\{Y_j\}$ are independent but the residuals are no longer exchangeable. However, under a randomization model the $\{Y_j\}$ and their corresponding residuals $\{e_j\}$ are considered fixed constants, and thus the permutation test remains valid. Also note that since the rank scores of the $\{Y_j\}$ are functions of the complete set of n responses, and thus can not be observed until the complete set has been randomized, so also the rank scores of the residuals $\{e_j\}$ can not be observed until the complete set is observed.

Another advantage of the randomization analysis of the residuals is that the validity of the test in no way depends on the validity of the model assumptions used to fit the model. Thus, if a simple normal errors model is used as the basis for computing the residuals, then the validity of a t or F-test between groups depends on the homoscedastic normal errors assumption. However, the permutation test comparing the randomly assigned groups in no way depends on this assumption. Thus the permutation test can be viewed as a robust test in situations where the regression model may be misspecified.

For a normal errors model, the simple residual is readily obtained as $e_j = (y_j - \hat{y}_j)$ where $\hat{y}_j = \hat{\alpha} + \mathbf{x}'_j\hat{\boldsymbol{\beta}}$ for the jth subject with covariate vector $\mathbf{x}_j$, as a function of the intercept α and coefficient vector $\boldsymbol{\beta}$. In a logistic regression model the $\{Y_j\}$ are binary indicator variables for a positive (1) versus negative (0) outcome, with conditional expectation

$$\pi_j = (1 + \exp[-(\alpha + \mathbf{x}'_j\boldsymbol{\beta})])^{-1}.$$

Then the simple residual is $e_j = (y_j - \hat{\pi}_j)$. The Pearson residual, defined as the root of the contribution to the Pearson chi-square goodness of fit, would be computed as

$$e_j = \frac{y_j - \hat{\pi}_j}{[\hat{\pi}_j(1 - \hat{\pi}_j)]^{1/2}}. \tag{8.38}$$

The logistic regression model is a member of the family of generalized linear models based on a conditional error distribution that is a member of the exponential family. In these models the conditional expectation $\mu(\boldsymbol{x})$ of $Y|\boldsymbol{x}$ is expressed as a linear function of the covariates through a link function such that $g[\mu(\boldsymbol{x})] = \alpha + \mathbf{x}'_j\boldsymbol{\beta}$ and $\mu(\boldsymbol{x}) = g^{-1}[\alpha + \mathbf{x}'_j\boldsymbol{\beta}]$. Then the error distribution specifies that the conditional variance is some function of the mean, expressed as $\sigma^2[\mu(\boldsymbol{x})]$. The corresponding Pearson residual is

$$e_j = \frac{y_j - \hat{\mu}(\mathbf{x_j})}{\hat{\sigma}[\mu(\mathbf{x_j})]}. \tag{8.39}$$

Alternately, the deviance residual can be computed as the contribution of the jth observation to the model deviance, the form of which is specific to each model. The expressions are given in many standard texts on linear models such as McCullagh and Nelder (1989).

For the analysis of survival data using a proportional hazards model, Therneau, Grambsch, and Fleming (1990) describe martingale residuals which examine the difference between the event counting process and the cumulative hazard which is the compensator for the process (cf. Fleming and Harrington, 1991). Let t_j refer to the observed time of death (or the event) for the jth subject, in which case $\delta_j = 1$, or the time up to which the subject was known to be alive (or event free) in which case $\delta_j = 0$. Then the martingale residual, defined at time t_j, is computed as

$$M(t_j) = \delta_j - \int_0^{t_j} \lambda_0(s)e^{\mathbf{x}'_j\boldsymbol{\beta}}ds,$$

where $\lambda_0(s)$ is the background hazard function. In this model $\lambda_0(s)$ replaces the intercept in the preceding linear models and the function must be estimated in order to compute the martingale residuals. A positive residual represents an individual who "died too soon" and a negative residual one "who survived too long" relative to others with the same covariate values. The deviance residual is a transformation of the martingale residual (Fleming and Harrington, 1991).

In the case of a blocked randomization, one approach to a regression model adjustment would be to simply add a block indicator covariate to the regression model. However, this leads to the inclusion of a large number of nuisance parameters in the model. An alternative would be to use a conditional regression model, such as a conditional logistic regression model or a stratified proportional hazard model, conditioning or stratifying on block. In a population model analysis, the model would only include treatment group and the adjusting covariates. In a randomization model analysis, one would fit the model using the adjusting covariates, compute the appropriate residuals, and then conduct a randomization analysis of the residuals.

8.8 EXAMPLE 1: THE NEONATAL INHALED NITRIC OXIDE STUDY

8.8.1 A Blocked Randomization and Analysis

The Neonatal Inhaled Nitric Oxide Study (Neonatal Inhaled Nitric Oxide Study Group, 1997) employed a permuted-block randomization and a random allocation rule within blocks to assign neonates requiring oxygen to receive either inhaled nitric oxide or placebo. The goal was to assess the effects of therapy on the need to apply extracorporeal membrane oxygenation (ECMO) or death, a combined primary outcome. The published analysis employed a simple population model analysis with a test for two proportions. Within 120 days following randomization, 64% of the control group versus 46% of the nitric oxide group died or required ECMO, $p < 0.006$. Herein we present a randomization-based analysis. The data from this study are available from the authors.

A total of 235 neonates were randomized (114 treated, 121 control) within 85 blocks of random length 2 or 4, with the exception of some Canadian clinics in which blocks of length 6 were also employed. Of these, 11 blocks contained subjects assigned to only one of the two treatments (4 treated, 7 control) and thus do not contribute to the blocked analysis. The unblocked analysis yields a Mantel-Haenszel chi-square test value $X_U^2 = 8.48$ ($p < 0.0037$). The blocked analysis using a separate stratum for each block yields a stratified-adjusted Mantel-Haenszel chi-square test value $X_B^2 = 8.54$ ($p < 0.0036$). This is equivalent to the stratified linear rank test in (8.10) using binary scores ($a_{jn} = 1$ or 0) with unit weights ($\omega_i = 1$) and allowing for incomplete blocks. The fact that the two test values are virtually identical indicates that the intrablock correlation is virtually zero. Thus subjects randomized within a block tend to be no more alike than subjects entered in different blocks.

For reference, the weighted test using the MVLE-like weights in (8.12) yields a chi-square value $W_B^2 = 6.40$ ($p < 0.0115$), and that using the stochastic ordering-like weights in (8.13) yields $W_B^2 = 7.99$ ($p < 0.0048$). This should not be taken to indicate that unit weights will in general provide a larger test value. The Mantel-Haenszel test is known to be optimal, in the sense of asymptotic efficiency, when there is a common odds ratio among strata in the population, or when the data show random variation about a common value. When the data arise from a population with a common relative risk then a different set of weights provides an optimal test. See Lachin (2000) for a discussion of these issues. While we present different weighted tests for illustration herein, in practice it is necessary that a single test, and weighting procedure, be specified beforehand.

The partial pressure of arterial oxygen (PaO$_2$) was a secondary outcome. Values were missing from 8 subjects (2 treated, 6 control). Nevertheless, the analysis is readily conducted under the independence assumption discussed in Section 8.5. The unblocked conditional test using rank scores yields a chi-square test value $X_U^2 = 22.83$ ($p < 0.0001$). In a blocked analysis, 15 blocks contained subjects assigned to only one of the two treatments and did not contribute to the analysis. The remaining 70 blocks contained 211 subjects, 106 assigned to the treated group and 105 to control. The unweighted blocked analysis using a separate stratum for each block yields a stratified-adjusted chi-square test value $W_B^2 = 9.07$ ($p < 0.0026$). The test using the MVLE-like weighted test yields $W_B^2 = 9.61$ ($p < 0.0020$), and that using the stochastic ordering-like weights yields $W_B^2 = 10.07$ ($p < 0.0016$). In this case, the fact that the blocked analyses yield smaller test values than the unblocked analysis indicates that the intrablock correlation of these measures is negative.

8.8.2 A Post-Stratified Blocked Analysis

As an illustration of a post-stratified blocked analysis, consider the analysis of the PaO$_2$ levels, stratified by an Apgar score ≤ 2 versus a score > 2. Due to the small block sizes (2, 4 or 6), when the blocks are further stratified by Apgar score, a total of 49 stratified blocks do not contribute to the analysis, constituting 52 observations, including 27 in the treated group and 25 in the control group. Thus one reason why this stratified-adjusted analysis differs from the analyses presented previously is the difference in sample sizes.

Table 8.4 presents the p-value for the blocked test using rank scores within Apgar score strata, and combined, using different weighted tests. Within the Apgar ≤ 2 subgroup, the tests have equivalent results. However, within the Apgar > 2 stratum, and combined, the MVLE-weighted test provides a somewhat larger p-value.

A similar post-stratified adjusted analysis can be readily applied to the primary outcome. Creating blocks defined jointly by the Apgar score strata and the randomization blocks within center, the jointly stratified/blocked-adjusted Mantel Haenszel chi-square test value is $W_B^2 = 7.096$ ($p < 0.0077$), slightly less significant than the analysis not stratified by Apgar score. This is equivalent to a stratified-combined blocked linear rank test with unit weights.

Table 8.4   *Post-stratified analysis of the Neonatal Inhaled Nitric Oxide Study using different weighting schemes.*

Subgroup	n_i	n_{iA}	$\omega_i = 1$	MVLE	Stochastic Ordering
Apgar ≤ 2	29	14	0.1206	0.1199	0.1152
Apgar > 2	146	71	0.0033	0.0114	0.0049
Combined	175	85	0.00097	0.0121	0.0030

8.8.3   Covariate-Adjusted Blocked Analysis

A further analysis was conducted using regression models to adjust for the influence of baseline covariates. For the analysis of the primary outcome, a binary variable, a logistic regression was fit with the baseline level of PaO_2, birth weight, gender and Apgar score. The model entropy R^2 is 0.093. Of the original 85 blocks, 17 blocks containing 19 subjects (6 treated, 13 control) were eliminated because they contained subjects assigned to only one of the treatment groups, leaving 61 blocks containing 187 subjects (91 treated, 96 control). A rank analysis was then applied to the Pearson residuals (8.38). The unweighted blocked analysis chi-square test yields $W_B^2 = 6.33$ ($p < 0.012$). The test using the MVLE-like weights yields $W_B^2 = 6.92$ ($p < 0.0086$), and that weighted by the stochastic ordering-like weights yields $W_B^2 = 7.17$ ($p < 0.0075$). All p-values are much smaller than the unadjusted blocked analysis.

For the analysis of PaO_2 a regression model adjusted for the baseline level of PaO_2, birth weight, gender and Apgar score explained 22 percent of the variation in the post-treatment level of PaO_2. In addition to the 8 subjects with a missing post-treatment PaO_2, an additional 27 subjects were eliminated from the analysis due to missing values of the covariates. Further, 20 blocks containing 22 subjects (8 treated, 14 control) were eliminated because they contained subjects assigned to only one of the treatment groups. The rank analysis of the residuals from the regression model adjustment was then applied to the remaining 61 blocks containing 178 subjects (88 treated, 90 control). The unweighted blocked analysis chi-square test yields $W_B^2 = 21.06$ ($p < 0.0000045$). The test using the MVLE-like weights yields $W_B^2 = 21.40$ ($p < 0.0000038$), and that weighted by the stochastic ordering-like weights yields $W_B^2 = 22.93$ ($p < 0.0000017$). All p-values are much smaller than the unadjusted blocked analysis.

Table 8.5 *Sample sizes, numerators, denominators, and Z-test results by stratum for the DCCT unconditional and conditional analyses.*

			Unconditional			Conditional		
i	n_i	n_{Ai}	S_i	V_i	Z_i	S_i	V_i	Z_i
1	31	17	-33.5	612.1	-1.354	-1.211	0.984	-1.221
2	22	13	-14.5	218.1	-0.983	-0.776	0.978	-0.785
3	29	15	34.5	499.8	1.543	1.511	0.998	1.513
4	27	15	-26.0	404.8	-1.292	-1.272	1.000	-1.272
5	28	11	-1.5	452.4	-0.071	-0.074	1.000	-0.074
6	29	14	-9.0	499.4	-0.403	-0.378	0.998	-0.379
7	27	15	14.5	402.6	0.723	0.683	0.999	0.683
8	23	11	-17.5	251.1	-1.104	-1.080	0.995	-1.082
9	29	14	-12.0	506.1	-0.533	-0.497	0.995	-0.499
10	28	11	23.0	445.6	1.090	1.155	0.997	1.157
11	32	18	6.0	679.9	0.230	0.233	1.000	0.233
12	23	8	-32.5	248.0	-2.064	-1.823	0.989	-1.833
13	15	7	1.5	68.4	0.181	0.274	0.951	0.281
14	18	9	7.0	120.6	0.637	0.602	0.996	0.603

8.9   EXAMPLE 2: THE DIABETES CONTROL AND COMPLICATIONS TRIAL

8.9.1   A Stratified Urn Randomization and Analysis

As described subsequently in Section 9.8.3, the Diabetes Control and Complications Trial (DCCT) employed Wei's urn design with $\alpha = 0$, $\beta = 1$ to randomize subjects to receive either intensive or conventional therapy within each of 25 clinical centers. The randomization was also stratified by primary versus secondary cohorts defined by the absence or presence, respectively, of pre-existing complications. One of the principal outcomes of the DCCT was the level of albumin excretion rate (AER) which is an indication of the extent of nephropathy (Diabetes Control and Complications Trial Research Group, 1995). For these analyses a random subset of 7 clinics is employed containing 361 subjects, 178 treated intensively and 183 conventionally. Complete data were available in the cohort at year 1, but at subsequent years there were missing values, principally due to staggered entry leading to administrative censoring. While the actual randomization was a bit more complicated than a simple $UD(0, 1)$, herein we present analyses assuming that a simple $UD(0, 1)$ had been employed within each stratum (i). These data are available from the authors.

Table 8.5 presents the linear rank test using simple rank scores within each of the 14 strata defined on the basis of the clinical center and primary versus secondary

Table 8.6 Overall tests using different MVLE-like weights and stochastic ordering (SO) weights for results in Table 8.5.

Analysis	Weights	S	V	W_B^2	$p <$
Unconditional	MVLE	-1.113	1.662	0.745	0.389
	SO	-10.291	123.7	0.856	0.355
Conditional	MVLE	-15.963	582.5	0.437	0.509
	SO	-0.443	0.398	0.494	0.483

cohort. The test statistic, variance and Z-value are presented for the unconditional analysis and the conditional analysis within each stratum. The unconditional and conditional Z-values are similar in all strata.

Of course we are principally interested in an overall test of the treatment effect, rather than the tests within strata. Table 8.6 presents the stratified-combined statistic (numerator of (8.8)) and variance, the test and Z-value using the MVLE-like weights and using the stochastic ordering (SO) weights. The tests using either the $MVLE$ or stochastic ordering weights are virtually identical in this analysis. Further, the unconditional test Z-value is slightly greater than the conditional test for these data.

8.9.2 Urn Analysis with Missing Data

The year 1 AER values were employed in the preceding analyses because there are no missing values. However, at subsequent years some subject values were missing, principally due to administrative censoring. Among this subset of clinics, the AER at 5 years was observed in 305 of the 361 subjects (148 intensive, 157 conventional), and missing in 56 subjects. Nevertheless, it is straightforward to conduct a randomization analysis of these data. Essentially the non-missing data comprise a subgroup among those entered. Unconditionally the analysis can be conducted using the simple complete randomization variance as described in Section 8.5.4. Since there is only one subgroup of interest within each clinic (those not missing), the unconditional variance is simply the sum of squares of the centered scores divided by 4. The test using the MVLE-like weights yields $W_B^2 = 5.30$ ($p < 0.022$), and that weighted by the stochastic ordering-like weights yields $W_B^2 = 4.93$ ($p < 0.027$).

The post-stratified conditional linear rank test for an urn design has not been studied.

8.9.3 Covariate-Adjusted Urn Analysis

A further analysis of the AER at 1 year was conducted to adjust for baseline covariates. A linear regression model was fit to the log(AER) at 1 year using covariates: log(AER) at baseline, gender, duration of diabetes, body mass index, mean blood pressure,

HbA$_{1c}$ (a measure of blood glucose exposure) and c-peptide (a measure of residual insulin secretion). The model accounted for 24 percent of the variation in the year 1 values. The adjusted unconditional analysis yielded results that were similar to the unadjusted analyses presented previously. For the unconditional analysis, the stratified-combined statistic using the MVLE-like weights is $W_B^2 = 0.419$ ($p < 0.518$). The stratified-combined statistic using the stochastic ordering weights is $W_B^2 = 0.540$ ($p < 0.463$). Again the tests are similar.

These adjusted analyses yield a larger p-value than the unadjusted analyses, 0.389 and 0.376, respectively. A similar effect is observed in population model-based normal errors analyses wherein the unadjusted group effect yields $p < 0.737$, whereas the adjusted effect yields $p < 0.846$. This is due to slight covariate imbalances between groups at baseline.

8.10   CONCLUSIONS

The conditions under which the randomization-based analysis will in general yield p-values less than population model based analyses, whether adjusted or not, has not been explored. If a randomization analysis is to be conducted, then in order to preserve the type I error rate at the desired level, one should analyze *exactly* as one randomized. If a permuted-block randomization was employed, this entails an analysis stratified by block. In this case, small block sizes, as in the nitric oxide study, will lead to exclusion of blocks that only contain subjects assigned to one of the two groups.

For a randomization-based analysis, pre-stratification has several advantages. First, one can discard strata based on *a priori* operational criteria without affecting the randomization stream. This is particularly relevant for pre-stratification by clinic in multi-center clinical trials, where a clinic may later be discarded in the analysis, such as when a clinic's participation is terminated due to lack of recruitment. Second, pre-stratification allows for a very simple stratified analysis by simply summing the numerator and denominator of the test statistic over the independent strata, possibly with stratum-specific weights. However, it should be clear from the developments in Section 8.4 that pre-stratification does not result in any benefits in terms of relative efficiency for a stratified analysis in large sample clinical trials.

In some cases a post-stratified analysis may be desired for covariates not considered in the pre-stratification process. One option is to conduct such an analysis with a separate test of treatment effects within strata. For the urn design, this induces correlations across strata that complicate the analysis, but these complications are not insurmountable, as discussed in Section 8.5. The within-stratum tests can then be combined to provide an overall stratified-adjusted assessment of the difference between the treatment groups.

If the objective of post-stratification is to provide an adjusted assessment of treatment effect, a simpler approach is to conduct a randomization-based test in a modeling setting, as discussed in Section 8.7. This also has the advantage that a randomization-based test can be conducted after adjusting for multiple covariates, including quantitative covariates.

Every trial includes *confirmatory* analyses of primary, and often secondary, objectives. For such analyses it is important that the principal method of analysis be specified *a priori*, whether randomization-based or population-based, whether adjusted for other covariates, and, if so, how the covariates are to be selected and how the adjustment is to be performed. However, after a trial has been conducted, a variety of *exploratory* analyses are conducted to address objectives beyond those explicitly stated in the protocol. Such analyses could be performed either under a randomization model or under an invoked population model, and many statisticians would favor a population model for covariate-adjusted regression analyses when considering such hypotheses.

8.11   PROBLEMS

8.1 Consider binary response data, where $a_j = 1$ or 0.
a. Under complete randomization, show that the linear rank test is algebraically equivalent to the Mantel-Haenszel test for a 2×2 table with mean (8.1) and variance (8.2).
b. Under the random allocation rule where $n_A = n_B = n/2$, show that the linear rank test is algebraically equivalent to the Mantel-Haenszel test for a 2×2 table with mean $s/2$ and variance $s(n - s)/4(n - 1)$.

8.2 Derive (8.15) and (8.16).

8.3 Derive (8.27).

8.4 Consider a randomization pre-stratified on a factor G of I levels and post-stratified on factor F of L levels. Derive an aggregate test over the IL strata by constructing a vector S consisting of the IL stratum-specific statistics.

8.12   REFERENCES

BRESLOW, N. E. (1981). Odds ratio estimators when the data are sparse. *Biometrika* **68** 73–84.

DAVIS, C. S. (1989). Two-sample post-stratified or subgroup analysis with restricted randomization rules, *Communications in Statistics - Theory and Methods* **18** 367–378.

DIABETES CONTROL AND COMPLICATIONS TRIAL RESEARCH GROUP. (1995). The effect of intensive therapy on the development and progression of diabetic nephropathy in the Diabetes Control and Complications Trial. *Kidney International* **47** 1703-1720.

FLEISS, J. L. (1981). *Statistical Methods for Rates and Proportions.* Wiley, New York.

FLEISS, J. L. (1986). *The Design and Analysis of Clinical Experiments.* Wiley, New York.

FLEMING, T. R. AND HARRINGTON, D. P. (1991). *Counting Processes and Survival Analysis.* Wiley, New York.

FLYER, P. A. (1998). A comparison of conditional and uncondtional randomization tests for highly stratified designs. *Biometrics* **54** 1551–1559.

GRIZZLE, J. E. (1982). A note on stratifying versus complete random assignment in clinical trials. *Controlled Clinical Trials* **3** 365–368.

LACHIN, J. M. (1988). Properties of simple randomization in clinical trials. *Controlled Clinical Trials* **9** 312–326.

LACHIN, J. M. (1999). Worst rank score analysis with informatively missing observations in clinical trials. *Controlled Clinical Trials* **20** 408–422.

LACHIN, J. M. (2000). *Biostatistical Methods: The Assessment of Relative Risks.* Wiley, New York.

LITTLE, R. J. A. AND RUBIN, D. B. (1987). *Statistical Analysis with Missing Data.* Wiley, New York.

MANTEL, N. AND HAENSZEL, W. (1959). Statistical aspects of the analysis of data from retrospective studies of disease. *Journal of the National Cancer Institute* **22** 719–748.

MATTS, J. P. AND LACHIN, J. M. (1988). Properties of permuted-block randomization in clinical trials. *Controlled Clinical Trials* **9** 345–364.

MATTS, J. P. AND MCHUGH, R. B. (1978). Analysis of accrual randomized clinical trials with balanced groups in strata. *Journal of Chronic Diseases* **31** 725–740.

MCCULLAGH, P. AND NELDER, J. A. (1989). *Generalized Linear Models.* Chapman and Hall, London.

NEONATAL INHALED NITRIC OXIDE STUDY GROUP. (1997). Inhaled nitric oxide in full-term and nearly full-term infants with hypoxic respiration failure. *New England Journal of Medicine* **336** 597-604.

PURI, M. L. (1965). On the combination of independent two-sample tests of a general class. *International Statistical Review* **33** 229–241.

THERNEAU, T. M., GRAMBSCH, P. M. AND FLEMING, T. R. (1990). Martingale hazards regression models and the analysis of censored survival data. *Biometrika* **77** 147–160.

VAN ELTEREN, P. H. (1960). On the combination of independent two-sample tests of Wilcoxon. *Bulletin of the International Statistical Institute* **37** 351–361.

WEI, L. J. AND LACHIN J. M. (1984). Two-sample asymptotically distribution-free tests for incomplete multivariate observations. *Journal of the American Statistical Association* **79** 653–661.

WEI, L. J. AND LACHIN, J. M. (1988). Properties of the urn randomization in clinical trials. *Controlled Clinical Trials* **9** 345–364.

9

Randomization in Practice

9.1 INTRODUCTION

Each method of randomization has properties that are better suited to specific applications than others. Thus the choice of a randomization procedure and its implementation depend in part on the design features of the study. In this chapter, we outline the basic steps in determining an appropriate randomization procedure to use, generating the randomization sequence, and implementing the randomization in the clinical trial.

The paramount objective of randomization is to provide an unbiased comparison of the treatment groups. However, if not carefully implemented, the randomization procedure can be subverted, even in a double-masked study. Also, randomization procedures in an unmasked study are susceptible to varying degrees to subtle biases introduced by the investigators (selection bias), and to subtle biases introduced in any study by non-random sequential entry of subjects over time (accidental bias). Models for the assessment of these types of bias are presented in the preceding chapters.

Another objective of randomization is to permit a randomization-based analysis based on the exact or large sample probability distribution of a test statistic over the reference set of possible randomization permutations. Whereas power is the usual criterion for the selection of sample size, and its analogue relative efficiency is used to choose a most powerful test under a specific alternative, power as a statistical property only applies indirectly to permutation tests. Thus power or efficiency is not a useful criterion for distinguishing a permutation test based on different randomization procedures. Herein we principally consider the different randomization procedures from the perspective of the control of bias.

It is important to remember, as discussed in Chapter 2, that randomization alone is not sufficient to provide an unbiased comparison of groups. Two other criteria (Lachin, 2000a) are required to ensure that a study result is unbiased. The first is that missing data from any randomized subjects, if any, do not bias the comparison of groups. This can be achieved by an intent-to-treat design in which all randomized subjects are followed so long as alive, able, and consenting. Second, the outcome assessments must be obtained in an equivalent and unbiased manner for all patients. The latter is obtained by double-masking, or single-masking those who conduct the outcome assessments.

9.2 STRATIFICATION

The initial considerations in the design of a study stem from the study objectives. These should be stated in such a manner that specifies the target population, the number of treatment groups, the treatment regimens to be compared, the principal outcome measure to be used to compare the effects of treatment, the sample size per group, and the duration of follow-up. Given these, the first step in the implementation of randomization is to determine the number of strata, if any, to be employed in the randomization.

Methods of stratified analysis under a randomization model are discussed in Chapter 8. Standard methods may also be applied under a population model, such as the Mantel-Haenszel stratified analysis of multiple 2×2 tables (cf. Lachin, 2000b). In Chapter 8 we showed that, while stratification of the randomization on a covariate will promote balance of that covariate within each randomized group, the statistical gains in efficiency resulting from the stratified randomization are indeed small compared to a stratified analysis using the same strata but without a stratified randomization. Thus stratification must be justified on other grounds.

In most multi-center clinical trials, the greatest source of patient heterogeneity with respect to covariates for the particular disease under study is the clinical center in which the subjects are recruited. Further, there is often a wide range of numbers randomized within the various clinical centers. In addition, in some trials poor performing clinics may actually be dropped from the study and it would be desirable that the elimination of the patients from one clinic not affect the integrity of the randomization in the remaining clinics. For all these reasons it is generally recommended that the randomization be stratified by clinical center or randomization site. (See Section 4.2.)

It may also be argued that one should also stratify on other major covariates. The motivation is to ensure that a major imbalance in such a covariate does not occur by chance among the treatment groups. However, as shown in Chapter 5, the probability of treatment imbalances is small, and any chance imbalances that do occur can effectively be adjusted for in the analysis. This assumes, however, that such covariate imbalances are due to chance among the cohort initally randomized and that there is complete follow-up of the randomized cohort. However, imbalances among groups selected from among those randomized, such as due to incomplete follow-up, cannot be adjusted for, with or without initial stratified randomization, without special unverifiable assumptions.

When stratification is deemed necessary for a large number of covariates, or if the total strata sizes are small, a covariate-adaptive randomization procedure can be employed. The Pocock-Simon procedure described in Chapter 4 has the benefit of being fully randomized, but very little is known about its theoretical properties.

9.3 CHARACTERISTICS OF RANDOMIZATION PROCEDURES

In this section, we restrict consideration to the randomization procedures evaluated in Chapters 3, 5, and 6. As shown therein, the basic issue is a tradeoff between the desire to promote or guarantee balance in the numbers of treatment assignments versus the susceptibility to either selection bias in an unmasked study, or accidental covariate imbalances.

While Efron's maximum eigenvalue assessment of accidental bias is of academic interest, in practice it is likely not a useful concept because it quantifies the magnitude of a severe bias, not its risk. Perhaps the only method that would be disfavored by Efron's accidental bias criterion is the truncated binomial design, without blocks or strata, since for this design the maximum bias increases in n. However, our simulations show that when used in permuted blocks, the truncated binomial design yields a distribution of covariate imbalances comparable to those of the permuted block design with a random allocation rule within blocks. In our view, susceptibility to accidental bias is not the most useful criterion for choosing one randomization procedure over another.

Rather, in an unmasked study, or one with the potential for unmasking, the principal concern is the potential for introduction of selection bias. From our experience, it is human nature to try to arrange for a patient whom one feels is better suited to receive treatment $A(B)$ to be more likely to receive that treatment. This could be done, for example, by scheduling the randomization visit when one thinks it is more likely that the next assignment will be $A(B)$. Susceptibility to selection bias is a major concern.

9.3.1 Consideration of selection bias

In a double-masked study, there is no susceptibility to selection bias. The only remaining considerations are the extent to which balance should be promoted, and the implications for data analysis. Given the stratum sample sizes, one should evaluate the probability of imbalances within strata and in aggregate with each procedure under consideration. In so doing one should note that minor imbalances, such as 55 : 45 within strata or in aggregate, have little impact on the statistical properties of the study (although there may be ethical consequences). Although such imbalances may be of cosmetic concern, this alone should not require consideration of more aggressive restricted randomization procedures to ensure or promote balance. Other reasons, however, which may justify such an aggressive approach are when the agent is a new drug therapy for which it is desired to ensure that approximately half the subjects are assigned to the active therapy to accrue the required patient years of exposure for the assessment of side effects; or where the experimental therapy is very expensive or staff intensive and for budgetary reasons balance is desired. In such studies with a small sample size in total, or within strata, complete randomization or urn designs may not be acceptable because there is a modest probability of such minor imbalances. Efron's biased coin design tends to have the best balancing properties among those designs that do not force perfect balance.

In an unmasked study, the potential for selection bias becomes a major concern. As shown in Chapter 6, the more predictable the sequence of assignments, or the higher the correlation among successive assignments, the greater the potential for clinic site staff to influence the composition of the treatment groups by scheduling patients for randomization visits so as to try to "beat" the randomization. The procedure most susceptible to such bias is the permuted block design with small block sizes. Under the Blackwell-Hodges model, under a convergent guessing strategy, random permuted blocks has no effect on the potential for such bias. Under this model, the potential bias is a constant regardless of the current imbalance between groups, or the magnitude of the probability of the next assignment. Matts and Lachin (1988) explored an alternate model in which bias is introduced only when the block size is known and the investigator can predict with certainty, or identify those assignments in the tail of a block where the probability of assignment to A or B is 1. In this case, the use of variable block sizes will substantially reduce this potential for bias.

In practice, neither model for selection bias is particularly accurate. In our experience, the greater the imbalance, the greater the temptation to attempt to beat the randomization. This leads to a model like that proposed by Stigler (1969). However, most of the different randomization procedures have not been studied under such a model. Regardless it is clear that in an unmasked study, it is a temptation to try to beat the randomization, and during the course of a study, many clinic staff succumb. Thus it is prudent to avoid blocked randomization unless the stratum or aggregate sample sizes are small. Although the use of variable block sizes may not help avoid bias, it would not hurt.

Some additional advantage might be accrued by using a truncated binomial procedure rather than a random allocation rule to generate the assignments within each block. As shown in Chapter 6, sequences with lower potential for selection bias under the Blackwell-Hodges model have higher probability of occurrence using the truncated binomial than the random allocation rule. However, there are also the sequences with a longer tail of assignments with certainty; *e.g.*, the $AABB$ sequence. Although not studied, one would conclude that the truncated binomial design would have greater potential for selection bias under the Matts-Lachin model of predictions with certainty when the block size is known. Thus if the truncated binomial design is employed, use of variable block sizes would be prudent.

With moderate stratum or total sample sizes, however, the urn design is markedly preferred to the blocked designs due to the much lower susceptibility to selection bias. This approach has been used in a number of large-scale, unmasked multi-center trials with stratum sample sizes of 30 and higher including the Diabetes Control and Complications Trial (The Diabetes Control and Complications Trial Research Group,

1986) and The Diabetes Prevention Program (Diabetes Prevention Program Research Group, 1999).

9.3.2 Implications for analysis

In Chapters 7 and 8, we describe methods of analysis based on the randomization distribution of a family of linear rank statistics with respect to the probabilities associated with each of the multitude of possible permutations. Such a randomization-based analysis has the advantage that it requires no assumptions other than the fact that a specific randomization procedure was employed and all observations were obtained in an unbiased manner. This is markedly different from the usual methods of analysis such as a t-test for means or a chi-square test for proportions, which rely on the concept of sampling at random from a homogeneous population.

If it is planned that all analyses will be justified under population model assumptions, then the method of randomization is irrelevant to the choice of an analytic strategy or procedure. This is the approach most often taken in the analysis of a clinical trial, and usually without controversy. However, in many instances, the variance of a test statistic can be markedly larger under population model assumptions than under randomization-based assumptions, and the resulting p-value higher.

Of course, the two approaches are assessing different questions. The randomization analysis addresses the probability that a difference at least as large as that observed among the n subjects randomized into the trial could have occurred by chance under the null hypothesis. The population model analysis addresses the probability that such a difference could have been observed in samples of n_A and n_B subjects drawn at random from their respective populations. Thus the randomization analysis allows a conclusion about the effects of treatment among the n patients studied, whereas the population model analysis allows a conclusion, through confidence interval estimation, about the effects of the treatment in the general populations from which the n patients were drawn.

As we stated in Chapter 7, in our opinion both approaches have value. We think it prudent to specify in the protocol that a randomization analysis would be used to conduct a test of the treatment group effect on the primary outcome, and that a population model analysis would be used to estimate the effect within the general population.

9.4 CHOICE OF RANDOMIZATION PROCEDURE

Lachin, Matts and Wei (1988) present a review of the properties of each of these non-adaptive randomization procedures and recommendations for their applications. They also discuss the role of covariate-adaptive randomization. The following is a summary of the potential applications of each method.

9.4.1   Complete randomization

Complete randomization provides optimal protection against various experimental biases such as selection and accidental bias. It also provides a randomization analysis that is asymptotically equivalent to the usual population model analysis, in most situations. However, it also provides the greatest potential for an imbalance in the number of treatment group assignments. In a large study, this may be simply a cosmetic concern.

9.4.2   Forced-balance designs

The random allocation rule and the truncated binomial design each impose strict balance on the numbers of assignments to each group provided that the complete sequence of allocations is filled. However, there is a potential for sizable imbalances at some points during the course of the trial. In an unmasked study both are susceptible to selection bias, with the random allocation rule more susceptible under the Blackwell-Hodges model. The truncated binomial is most susceptible to Efron's accidental bias, the possibility for an extreme covariate imbalance increasing in n. The random allocation rule allows a simple large sample randomization analysis, but the truncated binomial design does not (see Chapter 14).

9.4.3   Permuted block design

The most common application of a forced-balance design is through the use of permuted blocks to ensure balance after each block of assignments is filled. Within each block either the random allocation rule or the truncated binomial design can be used. For small to moderate block sizes, simulations show that the two approaches are equally susceptible to covariate imbalances, no more so than other designs. However, in an unmasked study, the truncated binomial is less susceptible to selection bias under the Blackwell-Hodges model than is the random allocation rule. Thus, if permuted blocks are to be used in an unmasked study, the assignments should preferably be generated using a truncated binomial rather than random allocation rule. In either case, variable block lengths are also recommended to prevent investigators from discerning the nature of the assignments.

Because balanced assignments to the treatment groups is assured within complete blocks, this design is most attractive with small studies, or studies with many small strata. For a double-masked study this approach is fine. However, for an unmasked study, even with the truncated binomial and random block lengths, an additional strategy could be considered. Rather than conduct the randomization sequentially as each patient is recruited, all biases can be eliminated by waiting until a set of patients have been recruited to fill a block, and then conduct all assignments simultaneously. Such *block simultaneous randomization* completely eliminates the potential for selection bias in an unmased study.

The proper randomization analysis with a permuted block design is a little more complicated since the analysis must be blocked (see Chapter 8). Whereas it is tempting to ignore the blocking in the analysis, to do so may substantially sacrifice power. Matts and Lachin (1988) show that the ratio of common tests with and without blocking is a function of the intrablock correlation, the loss of power increasing as the correlation increases. For example, in the extreme case of a block size of 2, then in effect the proper analysis is a matched-pairs analysis. Ignoring the blocking thus provides the wrong analysis.

The permuted block design analysis also provides a unique approach to account for any bias introduced by incomplete data. As described in Chapter 8, when data are missing for some subjects, a randomization analysis among the observed subjects can be justified as a special case of a post-hoc subgroup analysis under the assumption that missing data arises due to an independent random process, analogous to "missing completely at random" (MCAR) in a population model. However, a permuted block design allows an unbiased analysis even when the independence assumption is not plausible. In this case, missing data within a block may bias the comparison of treatment assignments within that block, but not within other blocks with complete data. Thus, a complete block subset analysis will provide an unbiased assessment of the differences between treatment groups. If this type of analysis is planned, then this would suggest that small block sizes should be employed, to reduce the numbers of subjects eliminated from the analysis due to missing data from another subject in that same block.

9.4.4   Biased coin-type designs

Efron's biased coin design, Wei's urn design, and generalizations, provide randomization sequences that control the likelihood of treatment imbalances without imposing strict balance. As such, they are less predictable than the blocked designs, and less susceptible to selection bias in an unmasked study. The urn design promotes balance early in the sequence of assignments, but approaches complete randomization as the sample size increases. Thus these designs are attractive in an unmasked study with a moderate or large sample size in total, or within strata. A conditional large-sample randomization-based analysis is a little more tedious to compute with Wei's urn design, but the analysis is readily programmed. Large-sample properties of randomization tests following Efron's biased coin design have not yet been established (see Chapter 14).

9.5   GENERATION AND CHECKING OF SEQUENCES

Standard computer packages such as SAS and S-Plus have built in random number generators that can be used to prepare the randomization sequence for a clinical trial. Using the notation of Chapter 3, a loop over the n patients will generate the assignments plus increment the function $N_A(j), j = 1, ..., n$. For all restricted randomization rules described, the probability of assignment to A, given by $p_{Aj} = E(T_j|\mathcal{F}_{j-1})$, is simply a function of $N_A(j-1)$. For each patient, a uniform random

```
data urn;
nb=0;
do j=1 to 50;
  x=ranuni(seed);
  if j=1 then p=1/2;
  else p=nb/(j-1);
  if x < p then t='A';
  else t='B';
  if t='B' then nb+1;
  output;
end;
run;
proc print;
  var j t;
run;
```

Fig. 9.1 *SAS code to generate a randomization sequence for* $n = 50$ *using the* $UD(0,1)$ *randomization procedure.*

number $U_j \in [0,1]$ is generated, and the following rule is applied:

$$\text{If } U_j \; \leq \; p_{Aj}, \text{ assign treatment } A$$
$$U_j \; > \; p_{Aj}, \text{ assign treatment } B.$$

It should be noted that randomization sequences are only as good as the random number generator. One should test any random number generator used with appropriate goodness-of-fit statistics. Some popular tests for random number generators are given, for example, in Law and Kelton (1982, Sec. 6.4) and Rukhin, Soto, Nechvatal, *et al.* (2000). The RANUNI function in SAS is highly regarded as a reliable random number generator. Figure 9.1 gives the SAS code to generate a sequence of 50 treatment assignments from Wei's $UD(0,1)$ procedure.

For the case of three groups, say A, B and C, this approach generalizes by determining the desired probability of each treatment allocation. Let the probability that patient j will be assigned to A be denoted p_{Aj} and the probability that patient j will be assigned to B be denoted p_{Bj}. Based on the random uniform number U_j, the jth subject is then assigned as follows:

$$A \quad \text{if} \quad U_j \leq p_{Aj};$$
$$B \quad \text{if} \quad p_{Aj} < U_j \leq (p_{Aj} + p_{Bj});$$
$$C \quad \text{if} \quad U_j > (p_{Aj} + p_{Bj}).$$

This approach immediately generalizes to the case of any number of treatments.

A random number should be specified as the seed to initialize the sequence. The random number seed and the edition/version of the random number generator should be documented so that the sequence can be regenerated and replicated if need be.

One frequent question from experimenters is whether one need actually draw a random permutation as the basis for the randomization, as opposed to simply writing down an "attractive" sequence of assignments, or using a systematic sequence such as $ABAB\cdots$. While the answer should be clear from the preceding chapters, it is helpful to explore the historical views on this. It has long been recognized that such "attractive" sequences are likely to be somewhat systematic, if not formally so, and that systematic sequences are susceptible to systematic biases. That is, the characteristics of the sample units differ in a systematic way which corresponds to the systematic differences in assignments, resulting in substantial bias. More importantly, due to their non-randomness, such designs often fail to provide an accurate measure of residual error, or an accurate reflection of the unexplained random variation. In R.A. Fisher's 1935 treatise *The Design of Experiments* that many consider the basis for formal randomization in experimentation, he first points out (p. 63) that

> ... the results of using arrangements which differ from the random arrangement ... are thus in one way or another undesirable since they will tend to underestimate or overestimate the true residual error.

Then in a discussion of a sytematic versus random Latin squares, he states (p. 77) that

> The failure of systematic arrangements came not from recognizing that the function of the experiment was not only to make an unbiased comparison, but to supply at the same time a valid test of its significance. This is vitiated equally whether the components affecting the comparisons are larger or smaller than those on which the estimate of error is based.

Despite the desire to achieve true "randomness", it is common practice that randomization sequences are examined and perhaps rejected and replaced if the sequence is considered undesirable. If the only consideration were the cosmetic properties of a sequence, then such rejection and re-randomization would be warranted. However, this practice violates the assumptions required for a randomization-based analysis which is based on the probabilistic structure over the complete reference set of possible permutations. If classes of random assignments could be pre-specified that are acceptable for cosmetic reasons, and those unacceptable, then the proper plan would be to state those specifications *a priori*, generate a single sequence of assignments that is acceptable, and then use the reference set of acceptable permutations as the basis for an inference. This is the approach taken by Berger, Ivanova, and Knoll (2002). However, the theory to support large-sample inference from a restricted reference set of permutations does not exist.

In his 1958 text *Planning of Experiments*, D. R. Cox states (p. 87), in reference to a table of random numbers, or equivalently to a sequence of random assignments, that

> Randomness is a property of the table as a whole, thus we should talk about permutations produced by a random method, rather than random permutations... whether or not [any permutations are] legitimate random permutations is to be decided by the methods by which they were produced, and not by inspecting them as individuals... The best plan is, if possible, to decide which arrangements are to be rejected before randomization. It is difficult to give general advise about which arrangements to reject, but the best rule is to have no hesitation in rejecting any arrangement that seems on general common-sense grounds to be unsatisfactory.

In general, the two elements one might inspect in a sequence of assignments are the maximal imbalance between groups at any point in the sequence, either as an absolute or a percentage difference, and the maximal length of a run of assignments to one treatment. Neither is a concern with a permuted-block randomization, with modest or small block sizes, but either could be a concern with complete randomization, an unblocked random allocation rule or truncated binomial design, or with a biased coin or urn design.

If one is to inspect a sequence using either criteria, then the most appropriate procedure would be to prespecify the acceptable limits before generating a candidate sequence and then applying the criteria. The criteria should be specified independently of treatment assignment so that an excess of either A or B assignments, or a run of either A or B assignments, each of fixed magnitude, would lead to rejection of the sequence. In this way it can be argued the restrictions of the reference set of possible permutations is symmetric, meaning that for every AB sequence eliminated, the mirror BA sequence is eliminated. One would need to carefully consider the impact that such a rejection rule would have on inferential procedures under a randomization model. How to do this is not clear.

9.6 IMPLEMENTATION

In most instances the random assignments are generated prior to the start of recruitment into a study and then a system specified for the implementation of the randomization as the subjects are recruited. This system of randomization will also provide for the single- or double-masking of the assignments.

9.6.1 Packaging and labeling

For a double-masked pharmaceutical trial it necessary that a supply of placebo material be provided that is indistinguishable from the active agent with respect to appearance, consistency, touch, weight, taste, smell, etc. Patients or their companions are often tempted to open a capsule or crush a pill to see if they can detect the presence/absence of the active agent. During clinic visits patient may compare the weight of their medication supplies. The same also applies to clinic staff. Thus the placebo and active material should be as equivalent as possible in all respects other than the active agent.

In a clinical trial of a surgical or intervention procedure, masking is implemented by use of a sham procedure for those randomized to the control group. In the extreme case of a major surgical procedure, this would include a trip to the operating room with anesthetization, incision, and wound closure. Clearly in most cases this would be considered unethical. In other less extreme cases, such as a minor procedure under local anesthesia, or an infusion, or use of radiation, etc., a sham may be acceptable.

Randomization in pharmaceutical trials is implemented differently from intervention or surgical trials because of differences in the nature of the therapies. The simplest method for the implementation of a pharmaceutical trial is to employ randomization by lots or supplies. In this case the physician or pharmacist has two lots of material labeled 1 or 2. If a patient is assigned to A (or B) then a supply of medication is drawn from lot 1 (or 2) and provided to the patient, or the physician/practicioner, for administration. This approach is fine for an unmasked study. However, this is a poor approach for a masked study because if any one patient's treatment assignment is unmasked, due to an adverse event, or overdose, or whatever, then the entire study is unmasked. A variation would be to use a blocked-lot procedure, such as randomization by lots within clinical centers. In this case unmasking of an individual study subject would unmask the assignments within that block, but not necessarily unmask the entire study, provided that the lot 1 and 2 contents were varied within sites.

The most secure way of implementing a double-masked randomization for a pharmaceutical trial is to provide a unique supply of medication, pre-packaged and labeled, for each individual subject. This could be bottles of medication (or other containers) with pre-assigned patient numbers, or supply numbers. One approach is to assign each patient a unique randomization number at the time of randomization, and to have a prepared supply of medication for that randomization ready for dispensation or administration at the time of randomization. For example, in a multi-center trial, patient numbers might be assigned of the form $ccxxx$ where cc is the clinical site number $(01, 02, ...)$ and xxx is the patient number within clinical site $(001, 002, 003, ...)$. When the patient is randomized, and assigned a unique randomization number, a unique supply of study material is assigned to that patient. In this way emergency or inadvertent unmasking of a single patient has no impact on the integrity of the masking of other patients.

This system, however, requires that each study subject be identified by two study numbers. Prior to randomization subjects undergo a period of screeing to assess eligibility to enter the study, and perhaps even a trial period of treatment with a placebo to assess compliance with the medication regimen. In some cases patient must also be withdrawn from pre-existing medications or be stabilized using a specified regimen. Thus a study number is assigned at the initiation of screening and another number assigned at the time of randomization.

A variation on this approach is to simply have a set of study supplies packaged and labeled according to one numbering scheme and a system to assign a supply number to a patient at the time of randomization. In this case a patient might be assigned a study screening number at the time of the initial screening visit, and then a study supply number at the time of randomization. This technique is useful for covariate-adaptive randomization and response-adaptive randomization (to be discussed in Chapter 10), where the randomization sequence cannot be generated in advance, and depends on either the individual patient's characteristics or previous patient responses. However, the number of A and B supplies required is unknown in advance, and so there must be an oversupply of drug available.

For an unmasked trial pharmaceutical trial a pharmacist can administer the medication using a supply of the active material, those assigned to control receiving nothing. For a surgical or intervention trial, the clinic staff need only know whether a subject is assigned to the experimental or control arms. In such cases subjects are assigned a study number at the time of initial screening and then later assigned to treatment A or B.

9.6.2 The actual randomization

The treatment assignment can be conveyed to the clinical sites in a variety of ways. The oldest type of system, and in some respects the least favorable, is a system of sealed envelopes. Envelopes labeled by study patient number are distributed to the sites and as each successive patient is randomized, the next envelope is opened and the enclosed study supply, or intervention group is assigned to that patient. This system should never be used because it allows all the envelopes to be opened in advance, thus potentially unmasking the sites to the sequence of assignments and opening the study to extreme selection bias.

Rather a system should be implemented which guarantees that future assignments remain unknown. One approach is to employ central randomization where the clinical site contacts a central center to either verify the randomization in a double-masked study, or to provide the treatment assignment in an unmasked study. In either case it is advisable that there be central verification of the randomization process. It is recommended that prior to actual randomization the central office verify that the subject meets the entrance eligibility criteria (with no exclusions), has consented to randomization and full study participation, and is ready to administer the treatment immediately. Even in the case where a pre-packaged supply of medication or study material is ready for assignment in the clinic, it is advisable that central verification be employed. In studies without these checks, patients have been randomized who did not meet all eligibility criteria, who did not consent or who did not ever receive any study medication. This is inexcusable. In a simple pharmaceutical trial, the recommended procedure is that the clinical site call a central office, verify that the patient is eligible and consenting, then have that patient's supply of medication ready for administration. The physician then meets with the patient, opens the bottle, removes a pill, hands the patient a glass of water and asks the patient to swallow the pill. If the patient balks then the patient is not randomized into the study. In this case the bottle of medication can be returned to the pharmacy for destruction, but the same patient number (supply) is assigned to the next willing patient.

Such central verification can be provided by a telephone call to central staff or by a central computer facility. In the latter case, the program asks the site to answer a series of questions using the key pad and then verifies or provides the random assignment. Another approach is to provide an interactive web site for this purpose.

Finally there must also be some system for post-randomization emergency unmasking. In most pharmaceutical trials, separate patient supplies of medication are prepared by a central distribution pharmacy based on the randomization sequence generated by the study statistician. The pharmacy should have operating procedures to ensure that each treatment (*e.g.*, active versus placebo) is processed in a manner to ensure that the randomization is executed as specified. This could include random selection of the packaged material for testing by a simple technique such as a litmus test. The central pharmacy should also provide a 24-hour answering service to answer any questions about the contents of each medication supply, and all supplies should be labeled with information about this service. In some studies the pharmacy may be instructed to notify a study monitor, such as the study medical monitor, of the unmasking.

9.7 SPECIAL SITUATIONS

The above discussion applies to a simple two (or more) group study design. There are studies, however, that present additional considerations.

In some studies, two (or more) active agents are to be used, such as where a new agent is compared to an active control, often manufactured or supplied by a different company. In this case it is not possible to provide an identical formulation for the two agents such that the study material is identical with respect to appearance, taste, etc., for each group. Then in order to maintain masking, a *double placebo* approach is necessary. Supplies of each active agent and a placebo for each agent are prepared, and each subject is asked to take two pills, one from bottle A and another from bottle B. If the patient is assigned to receive active treatment A (or B) then the supply of medication labeled A (or B) contains the active agent and the supply for the other agent labeled B (or A) contains the placebo. The patient takes two pills at a time, one containing active agent, the other a placebo.

This technique might be necessary in a two-group, positive controlled effectiveness trial, or an equivalence trial. A generalization is a 2×2 factorial design where patients are assigned to receive either control, A alone, B alone, or A and B in combination. In this case the A and B bottles contain placebo A and placebo B, active A and placebo B, placebo A and active B, and active A and active B, respectively.

In some cases, the A and B supplies also differ in form. For example, the A therapy may be administered orally as a capsule and the B agent by infusion. Regardless, masking can be preserved by administration of the matching placebo of the other agent to patients in either group.

In some cases one may have a design with an untreated control group and two active therapy groups where each therapy has a different formulation, requiring a double placebo approach to maintain complete double-masking. However, if one of the agents is administered orally, and the other by infusion, then a complete double placebo implementation may not be ethical. For example suppose that m subjects are assigned to each of treatments A (active oral), B (active infusion), and C (neither). Patients and local Institutional Review Boards may object to the administration of a sham (placebo) infusion to the $2m$ subjects assigned to treatments A or C. However, incomplete double-masking may be obtained by randomly assigning half the control patients to receive the A placebo, and half to receive the sham B infusion. In this manner m subjects receive A and $m/2$ the A placebo, and m subjects receive B and $m/2$ the B sham. This would be implemented by a two-stage randomization. First patients are randomly assigned to receive treatments A, B or C. Then those assigned to receive C are randomly assigned to also receive either the A placebo or the sham B infusion. Note that there are only three groups for purposes of statistical analysis: the subjects who receive the A placebo and the sham B infusion combine to form the C control group.

Finally, randomization procedures may be implemented so as to "share" controls in multiple parallel protocols. For example, suppose that a study is launched using A versus A placebo. Later another study is initiated in an identical population using identical procedures to compare B versus B placebo. For the period while the two studies overlap, an incompletely double-masked, three-arm randomization could be employed as above, where half the subjects assigned to the "control" group are then randomly assigned to receive the A placebo and half the B placebo. However, all subjects assigned to group C during the period of overlap would be included in the analysis of the A versus control in the A study, and also in the analysis of the B versus control in the B study. Thus the subjects in group C are contained in the control group for both the A study and also the B study.

Consider the case where study A alone is recruiting over some period, followed by the simultaneous recruitment to the A and B studies during a second period, followed by the close of recruitment to study A and continued recruitment to study B during a third period. During the first and third periods, a simple two group randomization is employed. In the middle period a three group randomization is employed to A, B or C. In order to maintain total double-masking, half those assigned to receive C during this middle period could then be assigned to receive either the A placebo or B placebo via a supplemental randomization. This three-group randomization should be implemented in such a way that the parallel two-group randomizations are not affected.

For a permuted block design with m assignments to each treatment per block, where m may vary among blocks, then during the first and third periods balanced blocks of length $2m$ are employed, while during the second period blocks of length $3m$ are employed. Let n_A, n_B, and n_C represent the number of patients assigned to A, B, and C, respectively. For a biased coin design the two group randomization in the A study would assign to group A with probability p_A when $n_C > n_A$ and likewise in the B study to group B with probability p_B when $n_C > n_B$. For example, consider the case where $p_A = p_B = 2/3$ so that the biased coin allocations are in either a 2:1 or 1:2 ratio, depending on whether the excess allocations in the past are to control or active, respectively. Then during the second period when the allocations for the A and B studies are performed simultaneously, the possible settings and the corresponding allocation ratios are shown in Table 9.1. These allocations will preserve the desired 2:1 ratio for assignments to A versus C and for B versus C, but there is no control over A versus B imbalance during the period of joint allocations.

Table 9.1 Biased coin allocation ratios to A, B and C such that the probability of assignment to A is $p_A = 2/3$, and to B is $p_B = 2/3$, when there is an excess number of prior allocations to C.

Imbalance	$A : B : C$
$n_A > n_B > n_C$	$1 : 1 : 2$
$n_A > n_C > n_B$	$1 : 4 : 2$
$n_B > n_A > n_C$	$1 : 1 : 2$
$n_B > n_C > n_A$	$4 : 1 : 2$
$n_C > n_A > n_B$	$2 : 2 : 1$
$n_C > n_B > n_A$	$2 : 2 : 1$

A similar strategy can be employed for an urn $UD(\alpha, 1)$ design. Initially, for the first period where only the A study is recruiting, the urn contains α balls of type A and of type C. At the end of this period, at the start of the second period, let n_{1A} and n_{1C} refer to the numbers of allocations to A and to C, respectively. Thus the urn contains $\alpha + n_{1C}$ balls of type A, and $\alpha + n_{1A}$ balls of type C. Then to initialize the randomization to also include study B, $\alpha + n_{1A}$ balls of type B are also added to the urn. This equals the number of C balls in the urn so that B and C are allocated with equal probability on the next draw. After each draw, a ball is added to the urn for each of the two types other than that drawn. At the end of the second period, at the conclusion of recruitment to study A, the urn contains $\alpha + n_{1A} + n_{2A} + n_{2C}$ balls of type B, and $\alpha + n_{1A} + n_{2A} + n_{2B}$ balls of type C, where n_{2A}, n_{2B}, and n_{2C} are the numbers of allocations made to each treatment during the second period. At this point, all A balls are removed from the urn, as well as the excess B and C balls from the initial period and due to the A allocations during the second period, leaving $\alpha + n_{2C}$ balls of type B and $\alpha + n_{2B}$ balls of type C. These are the numbers of balls that would have been in the urn had one started with randomization only to B and C which produced n_{2B} allocations to B and n_{2C} to C. For a $UD(\alpha, \beta)$ design, the above "n"values would be multiplied by β.

This design would also tend to balance the A to B allocations during the second phase, in addition to balancing the A to C and B to C allocations. This approach would also be used to continue allocations in a multi-group study after one of the arms has been discontinued, such as due to adverse events, as illustrated in the Diabetes Prevention Program example below.

9.8 SOME EXAMPLES

9.8.1 The Optic Neuritis Treatment Trial

Fifteen clinical centers enrolled 457 patients using a permuted-block design with a separate sequence for each clinical center. Patients were randomly assigned to one of three treatment regimens: intravenous methylprednisolone, oral prednisone, or oral placebo. Whereas the patients in the oral-prednisone and placebo groups were not informed of their treatment assignments, those in the intravenous-methylprednisolone group were aware of their assignments. The primary outcome was the development of multiple sclerosis, there being a significant reduction in risk among those assigned to steroids. (See Beck, Cleary, Trobe, *et al.*, 1993.)

9.8.2 Vesnarinone in congestive heart failure

In a preliminary study of the drug vesnarinone in the treatment of congestive heart failure, two clinics each recruited 40 subjects who were assigned to receive double-masked vesnarinone versus placebo. A permuted block design with block size of two was used. The study showed a reduction in mortality among these 80 subjects. (See Feldman, Bristow, Parmley, *et al.*, 1993.)

9.8.3 The Diabetes Control and Complications Trial

The Diabetes Control and Complications Trial (DCCT) enrolled 1441 subjects with type 1 (juvenile) diabetes within 29 clinical centers who were randomly assigned to receive either intensive versus conventional therapy for the control of blood glucose levels. The study showed that the intensive group, which maintained lower levels of blood glucose, had significantly reduced risk of microvascular complications of diabetes (Diabetes Control and Complications Trial Research Group, 1986, 1993). The DCCT enrollment was conducted in two stages. In the initial feasibility stage a total of 278 patients were recruited in 23 clinics during 1983–1998. The randomization was stratified by adults versus adolescents within each clinical center, 46 strata total. Due to the small sample size per clinic, and the requirement that each clinic enroll at least four adolescents, an initial permuted block of four subjects were assigned within each stratum, followed by a $UD(0,1)$ randomization. In the second stage of recruitment from 1984-1989, six new clinics were added and the randomization was stratified by a primary prevention cohort (no pre-existing retinopathy) versus a secondary intervention cohort (some pre-existing mild retinopathy, among other differences) within clinical center, 58 strata total. To initialize the urns for each strata within the original 23 clinics, the 278 assignments were post-stratified by primary/secondary cohort and clinical center. Then the appropriate number of balls of each type were placed in the urn for each stratum. For example, if there were four intensive and three conventional patients within the primary cohort of a given clinic from the feasibility phase, then three intensive and four conventional balls were placed in the urn for that stratum. The sequences allowing for 50 subjects within each of the 58 strata were then generated with a specified seed. For the six new clinics the sequences started with zero balls in each of their 12 strata. Each sequence was inspected to ensure that there were no long runs of assignments to either treatment. In one stratum a run of 9 As was followed by a run of 7 Bs. One element from each run was randomly selected to be changed to the other treatment. Of the 1441 patients randomized into the study, in the primary prevention cohort 378 were assigned to receive conventional therapy, 348 intensive therapy; and in the secondary intervention cohort 352 were assigned to receive conventional therapy, 363 intensive therapy.

9.8.4 Captopril in diabetic nephropathy

Thirty clinical centers recruited 409 subjects with pre-existing diabetic nephropathy. Using a $UD(0, 1)$ procedure stratified by clinical center, a total of 207 were assigned to receive double-masked captopril and 202 to receive placebo. The risk of further progression of diabetic nephropathy was reduced by 48 percent with captopril. (See Lewis, Hunsicker, Bain, *et al.*, 1993.)

9.8.5 The Diabetes Prevention Program

In the Diabetes Prevention Program (DPP), a total of 27 clinical centers recruited 3234 adult subjects with impaired glucose tolerance who were followed to observe the incidence of type 2 diabetes. Using a $UD(0, 1)$ procedure stratified by clinical center, subjects were assigned to receive either lifestyle intervention aimed at weight loss through diet and exercise, or conventional lifestyle management plus the drug troglitazone, or conventional management plus the drug metformin, or conventional management plus placebo. The lifestyle treatment was unmasked. In order to maintain masking among the three medication treatment groups, a double placebo technique was employed where each subject took two pills daily containing either one of the active agents or placebo. The troglitazone arm was terminated due to adverse effects after 585 patients had been randomized to receive troglitazone. These patients were unmasked and their treatment terminated. The remaining subjects were then told only to take their assigned pills from the metformin bottle, half containing active agent, half placebo. At that point, the composition of the urn was modified to shift from a four arm randomization to a three arm randomization. All the troglitazone balls were removed from the urn, as well as the 585 balls of each other type added due to these prior troglitazone allocations. At the end of the study, 1079 subjects had been assigned to lifestyle therapy, 1073 to metformin and 1082 to metformin-placebo. The study showed that lifestyle intervention achieved approximately a 58 percent reduction in the risk of developing diabetes versus placebo, whereas metformin yields a 31 percent risk reduction. (See Diabetes Prevention Program Research Group, 1999.)

9.8.6 Adjuvant chemotherapy for locally invasive bladder cancer

In an ongoing multicenter clinical trial of patients with locally invasive bladder cancer, who have undergone a radical cystectomy with lymph node dissection, and whose tumors demonstrate p53 abnormalities, 190 patients are to be randomized to either adjuvant chemotherapy (95 patients) or routine follow-up (95 patients) after surgery. The primary outcome is time to recurrence. Because of the large number of stratification variables, it was decided that the Pocock-Simon procedure would be used (see Section 4.4.2). Stratification variables were age (dichotomized at 65 years), stage (dichotomous), grade (dichotomous), and p21 status (dichtomous). The value of p in (4.4) was set to 0.75 ($c^* = 1.25$). In the first 54 patients randomized, the overall balance has been quite good with no imbalances greater than 2. (Personal communication, Susan Groshen.)

9.9   PROBLEMS

9.1 Using each of the following procedures, generate a separate randomization sequence for 50 random allocations to two groups:

(i) complete randomization;
(ii) random allocation rule;
(iii) truncated binomial design;
(iv) permuted blocks with $M = 5$;
(v) Efron's biased coin with $p = 2/3$;
(vi) Wei's urn design with $\alpha = 0$ and $\beta = 1$;
(vii) Smith's procedure with $\rho = 5$.

Provide the sequence and a copy of your program for each of the five procedures.

9.2 For each of the examples in Section 9.8, discuss the properties of the randomization procedure employed with respect to the potential for selection bias accidental covariate imbalance or other biases, and the implications for a randomization versus population model analysis.

9.3 Would complete randomization, with or without stratification as appropriate, be an acceptable approach in the DCCT? Justify your answer.

9.4 In each of the following cases, describe the randomization procedure you would employ and justify your answer. Generate the procedure and describe the resulting sequence.

(i) An investigator is planning a phase II trial with only 100 patients recruited in 10 clinical centers, with a range of 8–12 expected per center. The study will be double-masked and employ an active treatment versus control.
(ii) Consider an equivalent study but where the treatments by nature must be administered in an unmasked manner.
(iii) Now consider a Phase III study where 1000 patients are to be recruited in 20

clinical centers, with a range of 30–70 within each center. The trial will be double-masked and employ an active treatment versus control.

(iv) Consider an equivalent study but where the treatments by nature must be administered in an unmasked manner.

(v) Now consider a Phase III study where 1000 patients are to be recruited in 50 clinical centers, with a range of 10–30 within each center. The trial will be double-masked and employ an active treatment versus control.

(vi) Consider an equivalent study but where the treatments by nature must be administered in an unmasked manner.

9.5 In the vesnarinone study, since the block size used was two, subjects were randomly assigned within pairs. Would a randomization using a larger block size be acceptable? Justify your answer.

9.6 In the captopril study in diabetic nephropathy, approximately 25 percent of subjects entered the study with significant loss of renal function as represented by a serum creatinine exceeding 1.5 mg/dL. A stratified analysis was planned among those with such high values and among those with lower values. Should the randomization have also been stratified by high versus low initial creatinine values? Justify your answer.

9.7 As is now mandated for all major studies launched by the National Institutes of Health, one of the objectives of the Diabetes Prevention Program was to assess the effects of treatment among ethnic subgroups, both genders, and the elderly. The study recruitment targets included the recruitment of 50 percent ethnic minorities, including African Americans, Native Americans, Asian-Americans, and Hispanics; both genders, and 20 percent of subjects of at least 60 years of age. Should the randomization have also been stratified by any of these factors? Justify your answer.

9.10 REFERENCES

BECK, R. W., CLEARY, P. A., TROBE, J. D., KAUFMAN, D. I., KUPER-SMITH, M. J., PATY, D. W., AND BROWN, C. H. (1993). The effect of corticosteroids for acute optic neuritis on the subsequent development of multiple sclerosis. The Optic Neuritis Study Group. *New England Journal of Medicine* **329** 1764–1749.

BERGER, V. W., IVANOVA, A., AND KNOLL, M. D. (2002). Enhancing allocation concealment through less restrictive randomization procedures, with application to an unmasked trial of paclitaxel and carboplatin for advanced stage IIIB/IV non-small cell lung cancer. Submitted.

COX, D. R. (1958). *Planning of Experiments.* New York, Wiley.

DIABETES CONTROL AND COMPLICATIONS TRIAL RESEARCH GROUP. (1986). The Diabetes Control and Complications Trial (DCCT): Design and methodological considerations for the feasibility phase. *Diabetes* **35** 530–545.

DIABETES CONTROL AND COMPLICATIONS TRIAL RESEARCH GROUP. (1993). The effect of intensive treatment of diabetes on the development and progression of long-term complications in insulin-dependent diabetes mellitus. *New England Journal of Medicine* **329** 977–986.

DIABETES PREVENTION PROGRAM RESEARCH GROUP. (1999). Design and methods for a clinical trial in the prevention of type 2 diabetes. *Diabetes Care* **22** 623–34.

FELDMAN, A. M., BRISTOW, M. R., PARMLEY, W. W., CARSON, P. E., PEPINE, C. J., GILBERT, E. M., STROBECK, J. E., HENDRIX, G. H., POWERS, E. R., BAIN, R. P., WHITE, B. G., THE VESNARINONE STUDY GROUP. (1993). Effects of Vesnarinone on Morbidity and Mortality in Patients with Heart Failure. *New England Journal of Medicine* **329** 149–155.

FISHER, R. A. (1935). *The Design of Experiments.* Oliver and Boyd Edinburgh.

LACHIN, J. M. (2000a). Statistical considerations in the intent-to-treat principle. *Controlled Clinical Trials*, **21** 167–189.

LACHIN, J. M. (2000b). *Biostatistical Methods: The Assessment of Relative Risks.* Wiley, New York.

LACHIN, J. M., MATTS, J. P. AND WEI, L. J. (1988). Randomization in clinical trials: Conclusions and recommendations. *Controlled Clinical Trials* **9** 365–374.

LAW, A. M. AND KELTON, W. D. (1982). *Simulation Modeling and Analysis.* McGraw-Hill, New York.

LEWIS, E. J., HUNSICKER, L. G., BAIN, R. P., AND ROHDE, R. D. (1993). The effect of angiotensin-converting-enzyme inhibition in diabetic nephropathy. *New England Journal of Medicine* **329** 1456–1462.

MATTS, J. P. AND LACHIN, J. M. (1988). Properties of permuted-block randomization in clinical trials. *Controlled Clinical Trials* **9** 327–344.

RUKHIN, A., SOTO, J., NECHVATAL, J., SMID, M., BARKER, E., LEIGH, S., LEVENSON, M., VANGEL, M., BANKS, D., HECKERT, A., DRAY, J., AND VO, S. (2000). *A Statistical Test Suite for Random and Pseudorandom Number Generators for Cryptographic Applications.* National Institute of Standards and Technology, Gaithersburg.

STIGLER, S. M. (1969). The use of random allocation for the control of selection bias. *Biometrika* **56** 553–560.

10

Response-Adaptive Randomization

10.1 INTRODUCTION

We now revisit a topic mentioned in Chapter 3: the reasons behind equal allocation to the experimental and control therapies. Recall that two reasons were given, namely, power is maximized, and equal allocation reflects the view of equipoise that must exist at the start of the trial. Let us examine these two arguments afresh.

First, power is determined by the information accrued in the clinical trial, and under the traditional concept of statistical information, this is directly related to the variance of the test. If responses to the treatments have equal variability, power will be maximized under equal allocation. If they do not, power will be maximized using unbalanced allocation, with more patients allocated to the more variable treatment. For binary response problems, variability is directly related to the treatment effectiveness, whereas it will not be related for normal responses. In the latter case, one might have some indication at the beginning of the trial that one treatment is more variable than the other, and sample sizes can begin unbalanced. As the trial progresses, estimates of the variability could be obtained that would indicate that unequal allocation would result in more power. In the former case, since we are in a state of equipoise (and are essentially operating under the null hypothesis), we would not have any cause to employ unequal allocation, but this again may change as we accrue information on the treatment effect. So the power issue is considerably more complex than the oft-heard statement "unequal allocation results in a loss of power."

Similarly, should our view of equipoise at the beginning of the trial be fixed throughout the course of the trial, or could we use accruing data to dynamically alter the allocation probabilities to favor the treatment performing best thus far? It would seem that patients would then benefit by having less allocations to an "inferior" treatment (or at least inferior based on the data accrued thus far).

These are important practical and ethical questions that have been prominent in the literature since the 1950's; so prominent has been the debate that it is surprising that the need for equal allocation is nearly unquestioned in clinical trials of today.

In this chapter, we deal with *response-adaptive randomization*, in which the probability of being assigned to a treatment is changed throughout the trial according to data which have already accrued about the treatment effect. The goal of response-adaptive randomization is to assign more patients to the "better" treatment. These techniques fall under the broad category of *adaptive designs* (distinguished from response-adaptive randomization, which refers to *randomized* adaptive designs). Adaptive designs are useful in many disciplines (*e.g.*, Flournoy and Rosenberger (1995)) and have been proposed for clinical trials for many decades. Initial adaptive designs for clinical trials arose from considerations of optimal decision theory, including bandit problems, and of sequential stopping boundaries, and most of these designs were deterministic. We briefly review these designs from a historical perspective. We then discuss techniques for response-adaptive randomization, which affords the protections offered by all randomized experiments.

10.2   HISTORICAL NOTES

Adaptive designs in the clinical trials context were first formulated as solutions to optimal decision-making questions: Which treatment is better? What sample size should be used before determining a "better" treatment to maximize the total number receiving the better treatment? How do we incorporate prior data or accruing data into these decisions? The preliminary ideas can be traced back to Thompson (1933) and Robbins (1952) and led to a flurry of work in the 1960s by Anscombe (1963), Colton (1963), Zelen (1969) and Cornfield, Halperin, and Greenhouse (1969), among others. Perhaps the simplest of these adaptive designs is the *play-the-winner rule* originally explored by Robbins (1952) and later by Zelen (1969), in which a success on one treatment results in the next patient's assignment to the same treatment, and a failure on one treatment results in the next patient's assignment to the opposite treatment.

10.2.1   Roots in bandit problems

Consider a slot machine with two arms and a payoff that is observed immediately. To maximize the total payoff, which arm does one choose to play each time? In the context of clinical trials, the arms are the two treatments, and we desire to optimize some single objective, such as the mean squared error of an estimate of the treatment effect or the expected number of treatment failures. Such optimization problems are called *bandit problems* (*cf.* Berry and Fristedt, 1985; Gittins, 1989; Hardwick, 1995) and were originally proposed by Thompson (1933) and Robbins (1952).

Ideally one would like to switch back and forth from one treatment to the other any number of times to obtain the optimal sequence of treatment assignments. This is an extremely difficult problem even in the binary response case, because in order to find the optimal sequence, we have to specify a treatment to be used in each of the 4^n possible paths in a clinical trial with sample size n (*i.e.*, at each allocation, we could have a success on A, failure on A, success on B, failure on B). These problems also involve unknown parameters, and Bayesian and minimax approaches have been employed; see Berry and Fristedt (1985, Chapter 1) for a review of these techniques.

Discrete bandit problems can be solved using *dynamic programming* (Bellman, 1956). In the past, dynamic programming algorithms for even moderate sample sizes were computationally infeasible, but the advent of parallel processing and faster workstations has allowed some researchers to begin exploring both the feasibility of using this approach and the properties of this approach. Much of the seminal work in developing computational algorithms has been done by Hardwick and Stout (*cf.* 1995, 1999).

The difficulty in finding the optimal sequence using dynamic programming led some researchers to find alternative adaptive allocation procedures (*e.g.*, Berry, 1978). Most of these procedures are *myopic strategies*, in which the allocation rule attempts to optimize the treatment assignment for the current patient, by allocating to the treatment that has the has performed best thus far in the trial. It is well known that myopic strategies are not necessarily globally optimal (Berry (1986, p. 4)). Bandit solutions have the advantage that they balance the myopic goal (the patient at hand) with future rewards.

Optimal sequences from bandit solutions are deterministic. There has been very little literature on randomized bandit solutions. Berry and Eick (1995, p. 232) suggest the following:

> Assignment bias can be avoided ... by introducing an unbalanced randomization in which the treatment opposite from the one assigned by the procedure is used with probability sufficiently great to ensure blindness but not so large that the advantage of the adaptive procedure is lost – perhaps this probability can be between $1/10$ and $1/3$.

Hardwick and Stout (personal communication) have recently begun incorporating randomization into dynamic programming equations and have found that the degree of randomization degrades the performance of the optimal strategy, but not quite linearly. We recently became aware of a paper that gives a detailed treatment of randomized multi-armed bandit problems (Yang and Zhu, 2002).

10.2.2 Roots in sequential stopping problems

The previous discussion involved a fixed sample size. Others have examined adaptive designs in the context of a random number of patients N, in conjunction with an appropriate stopping boundary. The early papers taking this approach were Chernoff and Roy (1965), Flehinger and Louis (1971), Robbins and Siegmund (1974), Louis (1975), and Hayre (1979), among others.

In the Robbins and Siegmund (1974) model, assume that responses $x_1, ..., x_m$ and $y_1, ..., y_n$ are realizations of random variables that are independent and identically distributed as $N(\mu_1, 1)$ and $N(\mu_2, 1)$, respectively, and it is desired to test the hypothesis $H_0 : \mu_1 > \mu_2$ versus $H_1 : \mu_1 < \mu_2$. After each response, we observe the test statistic

$$z_{m,n} = \frac{mn}{m+n}(\bar{y}_n - \bar{x}_m).$$

Let $b > 0$ be a constant; we stop the trial as soon as $z_{m,n} \notin (-b, b)$ and declare H_0 is true if $z_{m,n} < -b$ and H_1 is true if $z_{m,n} > b$. Under an appropriate choice of b, we have Wald's sequential probability ratio test with fixed error probabilities α and β. If we wish to minimize the expected number of observations on the treatment with the smaller mean (i.e., the expected number of patients on the inferior treatment), it is logical to assume that equal allocation is preferable when $z_{m,n}$ is close to 0, and when $z_{m,n}$ is close to b or $-b$, most observations should be taken from the x population or the y population, respectively. Robbins and Siegmund propose the following rule. Let $c \geq b$. Having observed $x_1, ..., x_m$ and $y_1, ..., y_n$, the next observation should be y_{n+1} if

$$\frac{n-m}{m+n} \leq \frac{z_{m,n}}{c};$$

otherwise, the next observation should be x_{m+1}. The authors give some guidelines as to the choice of c and conclude that the error probabilities are essentially independent of the sampling scheme.

Other rules, other response models, and other types of hypotheses are explored by Flehinger and Louis (1971), Louis (1975), and Coad (1991), among others.

As in the decision theory approach, most of these approaches to adaptive designs have used nonrandomized allocation rules. As Rosenberger (2002) points out,

> Surprisingly, the link between [response-adaptive randomization] and sequential analysis has been tenuous at best, and this is perhaps the logical place to search for open research topics.

10.2.3   Roots in randomization

Both the bandit and sequential approaches discussed are fully adaptive designs, in that they select future treatments on the basis of all past information about that treatment. However, they have generally been developed for deterministic allocation, and hence are subject to biases that may be present in nonrandomized studies. In particular, Bather (1995) found that, for both the Robbins and Siegmund procedure and other adaptive designs, "selection bias can have a substantial effect in distorting the results of comparative experiments" (p. 32). Selection bias, as discussed in Chapter 6, is a serious problem for nonrandomized studies. But much of the recent research in adaptive designs has involved fully randomized designs. Response-adaptive randomization alters the allocation probabilities to reflect the current trend of the data, so that patients are assigned to the most "successful" treatment with probability less than 1.

Wei and Durham (1978) were perhaps the first to discuss response-adaptive randomization, in their famous *randomized play-the-winner rule* paper. The rule can be described as follows. An urn contains α balls representing treatment A and α balls representing treatment B. A ball is drawn and replaced. If the ball was type $i = A, B$, treatment i is assigned. A success on one treatment results in the addition of β balls representing that treatment, for a positive integer β. A failure on one treatment results in the addition of β balls representing the opposite treatment. Hence, unlike Zelen's model, we skew the probability of assignment to favor the treatment performing "better" (*i.e.*, less failures/more successes), rather than switching deterministically between treatments. This design is usually designated $RPW(\alpha, \beta)$.

Urn models are only one approach to accomplish response-adaptive randomization. We discuss these varied approaches in the remaining portions of this chapter. Because this book is about randomization, we will future discussion will principally focus on fully randomized adaptive designs.

10.3 OPTIMAL ALLOCATION

In this chapter we will explore (1) response-adaptive randomization that is based on optimal allocation targets, where a specific criterion is optimized based on a population response model, and (2) design-driven response-adaptive randomization, where myopic rules are established that have intrinsic intuitive motivation and can be completely nonparametric, but are not optimal in a formal sense. Let us begin with determining optimal allocation targets. Because these targets typically depend on unknown parameters of a response-distribution, they cannot be implemented in practice without some form of estimation. Hardwick and Stout (1995) review several criteria that one may wish to optimize, including expected number of treatment failures, expected number of successes lost, expected number of patients assigned to the inferior treatment, the total expected sample size, the probability of correct selection, or total expected cost. When the goal is to maximize the experience of individual patients in a clinical trial, the first three criteria are often used. One can argue their relative merits; for example, expected number of treatment failures takes into account the randomness inherent in the response model in that some patients may not benefit from the superior treatment, whereas expected number of patients assigned to the inferior treatment ignores this randomness and focuses on what the scientist can actually control. Of course, the first two criteria, expected failures and expected successes lost, refer to binary response trials where we have a clearly defined "success" and "failure".

These optimal allocation rules are derived under simple homogeneous population models, in which responses of patients assigned to the same treatment are assumed to follow the same distribution. This may be an unreasonable assumption, for example when there are important covariates that effect response or time trends. But it provides a simple working model to explore properties of response-adaptive randomization. In Chapter 12 we discuss the heterogeneity issue from a practical standpoint.

The general optimization approach we will employ derives from the approach of Jennison and Turnbull (2000), and can be traced back to ideas of Hayre (1979). The idea is to fix the variance of the test statistic to be constant and then to find an optimal allocation ratio R^* from the possible values of $R = n_A/n_B$ according to our particular criterion. In Jennison and Turnbull's approach, let Y_{Ai} arise from a $N(\mu_A, \sigma_A^2)$ distribution and Y_{Bi} arise from a $N(\mu_B, \sigma_B^2)$ distribution, $i = 1, 2, ...,$ and σ_A^2 and σ_B^2 are known. Then the denominator of the usual Z-test is the square root of the variance of $\bar{Y}_A - \bar{Y}_B$, given by

$$\frac{\sigma_A^2}{n_A} + \frac{\sigma_B^2}{n_B},$$

and we set this equal to a constant, say K. Let $n = n_A + n_B$ be a fixed number of patients in a clinical trial. Then we can write $n_A = Rn/(1+R)$ and $n_B = n/(1+R)$, and we obtain

$$n = \frac{\sigma_A^2(1 + R) + \sigma_B^2 R(1 + R)}{KR}. \tag{10.1}$$

Let $\theta = \mu_A - \mu_B$ be the true treatment effect. We wish to find the value of R that minimizes

$$u(\theta)n_A + v(\theta)n_B, \tag{10.2}$$

where u and v are appropriately chosen functions of θ. Because we wish to put more patients on treatment A if $\theta > 0$ and more patients on treatment B if $\theta < 0$, Jennison and Turnbull explore functions where u and v are strictly positive, and $u(\theta)$ is decreasing in θ for $\theta < 0$ and $v(\theta)$ is increasing in θ for $\theta > 0$. See Jennison and Turnbull (2000, p. 328) for details on choosing these functions. Substituting (10.1) into (10.2), we obtain

$$\frac{u(\theta)(\sigma_A^2 + \sigma_B^2 R) + v(\theta)(\sigma_A^2/R + \sigma_B^2)}{K}.$$

Taking the derivative with respect to R and equating to zero, we achieve a minimum at

$$R^* = \frac{\sigma_A}{\sigma_B}\sqrt{\frac{v(\theta)}{u(\theta)}}. \tag{10.3}$$

An interesting case arises. If $u = v = 1$, then we have $R^* = \sigma_A/\sigma_B$, which is simply Neyman allocation (see Problem 2.6), and maximizes the power of the usual Z-test. Note that this formulation also presents an alternate, but equivalent, interpretation. When $u = v = 1$, (10.2) is finding the optimal allocation to minimize the total sample size for a fixed variance of the test.

The general formulation with binary response was considered by Hayre and Turnbull (1981) in the context of sequential estimation. Rosenberger, Stallard, and Ivanova, et al. (2001) also deal with binary responses. If we let the responses on treatment A follow a Bernoulli distribution with parameter p_A and the responses on treatment B follow a Bernoulli distribution with parameter p_B, we can formulate an optimality criterion as in (10.2), however now the variances depend on p_A and p_B. We also have a dilemma as to which measure of the treatment effect we wish to use. Let $q_A = 1-p_A$ and $q_B = 1-p_B$. We could take the simple difference, $\theta = p_A - p_B$, the relative risk of failure, $\theta = q_B/q_A$, or the odds ratio, $\theta = p_A q_B/p_B q_A$. In any event, if we wish to minimize the expected number of treatment failures, (10.2) can be written with $u(\theta) = q_A$ and $v(\theta) = q_B$. The simple difference measure is analogous to the difference of means in the normal case above, and hence we obtain

$$R^* = \sqrt{\frac{p_A q_A}{p_B q_B}} \sqrt{\frac{q_B}{q_A}} = \sqrt{\frac{p_A}{p_B}}, \tag{10.4}$$

by (10.3). This differs from Neyman allocation, given by

$$R^* = \sqrt{\frac{p_A q_A}{p_B q_B}}. \tag{10.5}$$

If we use the other measures, we obtain different allocations. Consider the relative risk measure q_B/q_A. We can use the delta method to write the asymptotic variance as

$$\frac{p_A q_B^2}{n_A q_A^3} + \frac{p_B q_B}{n_B q_A^2}.$$

Substituting $n_A = Rn/(1+R)$ and $n_B = n/(1+R)$ and equating to K, we obtain

$$n = \frac{p_A q_B^2 (1+R) + q_A p_B q_B R(1+R)}{q_A^3 RK}.$$

Then our optimization criterion becomes finding the value of R to minimize

$$n \frac{q_A R + q_B}{1+R} = \frac{p_A q_A q_B^2 R + q_A^2 p_B q_B R^2 + p_A q_B^3 + q_A p_B q_B^2 R}{q_A^3 RK}.$$

Taking the derivative with respect to R and equating to zero, we obtain

$$R^* = \sqrt{\frac{p_A}{p_B} \frac{q_B}{q_A}}.$$

Table 10.1 gives the asymptotic variances and optimal allocation for the three types of measures.

The selection of appropriate measure would normally be dictated by the choice of test statistic. The most common is the simple Z test based on the simple difference; see Lachin (2000, Problem 2.9) for asymptotically equivalent tests based on smooth functions of p_A and p_B. Rosenberger, Stallard, Ivanova, et al. (2001) discuss tests based on the pooled versus the separate variance estimators.

Table 10.1   Asymptotic variances and optimal allocation for minimizing expected number of failures at a fixed variance, for three measures of the treatment effect from binary response trials.

Measure	Asymptotic Variance	R^*
Simple difference	$\dfrac{p_A q_A}{n_A} + \dfrac{p_B q_B}{n_B}$	$\sqrt{\dfrac{p_A}{p_B}}$
Relative risk	$\dfrac{p_A q_B^2}{n_A q_A^3} + \dfrac{p_B q_B}{n_B q_A^2}$	$\sqrt{\dfrac{p_A}{p_B}}\dfrac{q_B}{q_A}$
Odds ratio	$\dfrac{p_A q_B^2}{n_A q_A^3 p_B^2} + \dfrac{p_A^2 q_B}{n_B q_A^2 p_B^3}$	$\sqrt{\dfrac{p_B}{p_A}}\dfrac{q_B}{q_A}$

10.4   RESPONSE-ADAPTIVE RANDOMIZATION TO TARGET R^*

Since the optimal allocation involves unknown parameters of the population model, we cannot implement it in practice, unless we implement it under the null hypothesis, resulting in equal allocation, or under some other "best guess" of the parameter values. In this section we discuss two sequential methods, the *sequential maximum likelihood procedure* and the *doubly-adaptive biased coin design*.

10.4.1   Sequential maximum likelihood procedure

As the trial progresses, perhaps the most logical approach would be to substitute current values of parameter estimates for the unknown parameters; *i.e.*, if $R^*(\theta)$ is a function of an unknown parameter θ, after $j - 1$ patients, substitute $\hat{\theta}(j - 1)$ for θ. Then we can impose the following allocation rule. Let $\mathcal{F}_n = (T_1, ..., T_n, Y_1, ..., Y_n)$, where $T_1, ..., T_n$ assume the value 1 if treatment A and 0 if treatment B, and $Y_1, ..., Y_n$ are the first n responses to treatment. Then, using similar notation as in Chapter 3, the allocation rule is given by

$$E(T_j | \mathcal{F}_{j-1}) = \frac{R^*(\hat{\theta}(j - 1))}{1 + R^*(\hat{\theta}(j - 1))}. \tag{10.6}$$

While the $\hat{\theta}$ can be any estimator, it is usual to employ the maximum likelihood estimator of the assumed population model. Then the allocation rule in (10.6) is called the *sequential maximum likelihood procedure*.

For example, let us assume the simple binomial model where $Y_i = 1$ if there is a treatment success and $Y_i = 0$ if there is a treatment failure, $i = 1, ..., n$. The allocation rule for minimizing the expected number of treatment failures under binary response with the simple difference measure, from (10.4), is given by

$$E(T_j | \mathcal{F}_{j-1}) = \frac{\sqrt{\hat{p}_A(j - 1)}}{\sqrt{\hat{p}_A(j - 1)} + \sqrt{\hat{p}_B(j - 1)}},$$

Table 10.2 *Simulated values of expected allocation proportions,* $E(N_A(n)/n)$ *(standard deviation), for the sequential maximum likelihood procedure targeting* (10.4) *(A), the sequential maximum likelihood procedure targeting Neyman allocation* (N), *and equal allocation* (E), 5000 *replications (Rosenberger, Stallard, Ivanova, et al., (2001, p. 911), reprinted with permission of International Biometric Society).*

p_A	p_B	n	A	N	E
0.1	0.2	526	0.42 (0.04)	0.43 (0.04)	0.50 (0.02)
0.1	0.3	162	0.39 (0.06)	0.42 (0.05)	0.50 (0.04)
0.1	0.4	82	0.38 (0.07)	0.42 (0.06)	0.50 (0.05)
0.4	0.6	254	0.45 (0.04)	0.50 (0.03)	0.50 (0.03)
0.6	0.9	82	0.45 (0.06)	0.58 (0.06)	0.50 (0.05)
0.7	0.9	162	0.47 (0.04)	0.58 (0.05)	0.50 (0.04)
0.8	0.9	526	0.48 (0.02)	0.57 (0.04)	0.50 (0.02)

where

$$\hat{p}_A(j-1) = \frac{\sum_{i=1}^{j-1} T_i Y_i}{\sum_{i=1}^{j-1} T_i} \text{ and } \hat{p}_B(j-1) = \frac{\sum_{i=1}^{j-1}(1-T_i)Y_i}{\sum_{i=1}^{j-1}(1-T_i)}.$$

Similarly we can define a sequential maximum likelihood procedure for Neyman allocation, from (10.5), using

$$E(T_j|\mathcal{F}_{j-1}) = \frac{\sqrt{\hat{p}_A(j-1)\hat{q}_A(j-1)}}{\sqrt{\hat{p}_A(j-1)\hat{q}_A(j-1)} + \sqrt{\hat{p}_B(j-1)\hat{q}_B(j-1)}},$$

where $\hat{q}_A(j-1) = 1 - \hat{p}_A(j-1)$ and $\hat{q}_B(j-1) = 1 - \hat{p}_B(j-1)$. Properties of the sequential maximum likelihood procedure targeting Neyman allocation are explored by Melfi and Page (1995) and Melfi, Page, and Geraldes (2001). Properties of the sequential maximum likelihood procedure targeting (10.4) are explored in Rosenberger, Stallard, Ivanova, et al. (2001).

In the latter paper, a simulation was conducted to compare equal allocation, sequential maximum likelihood procedure targeting Neyman allocation, and sequential maximum likelihood procedure targeting (10.4). Results are given in Table 10.2 (the sample sizes were selected to give approximately 90 percent power for the usual Z-test under equal allocation). One can see that the Neyman allocation places too many patients on the inferior treatment for large values of p_A and p_B (see also Problem 10.4). Note that the allocation rule for minimizing expected treatment failures does put fewer patients on the inferior treatment, it is more variable than equal allocation. In general, this will be the case with sequential maximum likelihood procedures, and the variability is induced by the correlation among the treatment assignments. We will reflect on this more when we discuss power in the next chapter.

In what way does the sequential maximum likelihood procedure target the optimal allocation? It seems intuitively reasonable, since maximum likelihood estimators are typically consistent, to assume that $R^*(\hat{\theta}(n)) \to R^*(\theta)$, and that

$$\lim_{n \to \infty} \frac{N_A(n)}{n} = \frac{R^*(\theta)}{1 + R^*(\theta)}, \tag{10.7}$$

and hence we attain optimal allocation in the limit. However, we no longer have independent data because of the response-adaptive randomization, and the proof is considerably more complicated. This issue is addressed in detail in Chapter 15. It turns out that, under very mild conditions, (10.7) is true, and hence the sequential maximum likelihood procedure is asymptotically optimal. The reader is referred to Chapter 15 for mathematical details.

For the difference of normal means, Jennison and Turnbull (2000) propose a group sequential adaptive design. Suppose there are K interim inspections of the data. At stage $k, k = 1, ..., K$, the optimal allocation ratio is determined from (10.3) by substituting the current estimates of θ, $\hat{\theta}(k - 1)$, from the previous stages into $u(\theta)$ and $v(\theta)$. Then the next group will have size $n_{Ak} + n_{Bk} = n_k$, where n_{Ak} and n_{Bk} are determined by the estimated optimal allocation proportions (to an integer approximation) and n_k is a function of the amount of information accrued. This is a deterministic rule; Jennison and Turnbull mention that one could set a minimum sample size n^* for both arms to preserve at least some "randomization" in the treatment allocation and maintain an element of masking. One could alternatively establish a sequential maximum likelihood procedure that randomizes patients one-by-one using the estimated optimal allocation as the allocation probability. In this case, one must estimate both the mean and variance, unless one assumes known variances, as do Jennison and Turnbull (2000).

10.4.2   Doubly-adaptive biased coin design

Eisele (1994) and Eisele and Woodroofe (1995) propose a more complicated design to achieve the desired allocation proportion R^*. They refer to this design as the *doubly-adaptive biased coin design*. Let t be a function from $[0, 1]^2$ to $[0, 1]$ such that the following four conditions hold: (i) t is jointly continuous; (ii) $t(a, a) = a$, (iii) $t(a, b)$ is strictly decreasing in a and strictly increasing in b; and (iv) t has bounded derivatives in both arguments. The function t will represent a measure of the difference between $N_A(j)/j$ and $R^*(\hat{\theta}(j))$. Then we allocate to treatment A with probability

$$E(T_j | \mathcal{F}_{j-1}) = t\left(\frac{N_A(j-1)}{j-1}, R^*(\hat{\theta}(j-1))\right).$$

The properties of this design will depend largely on the function t used. Eisele and Woodroofe (1995) show that (10.7) holds, but under somewhat restrictive conditions. Melfi, Page, and Geraldes (2001) point out that the example in the last section of Eisele (1994) does not satisfy the requisite conditions, so one must be careful to choose t carefully.

10.5   URN MODELS

The preceding optimal allocation rules are based on parametric models and response-adaptive randomization that targets the optimal allocation are based on maximum likelihood estimates. A *design-driven* approach to the problem has been developed on an independent track from the optimal allocation approach. The basic idea is to use an intuitive rule to adapt the allocation probabilities as each patient enters the trial. While these rules do not have any optimal properties, we can determine limiting properties of the design, which may be attractive in their own right. One approach involves *urn models*, which include the randomized play-the-winner rule as a special case; designs based on urn models are completely nonparametric.

10.5.1   The generalized Friedman's urn model

A typical urn model for response-adaptive randomization is the generalized Friedman's urn model [Athreya and Karlin (1968)]. Initially a vector $Y_1 = (Z_{11}, ..., Z_{1K})$ of balls of type $1, ..., K$ are placed in an urn. Patients sequentially enter the trial. When a patient is ready to be randomized, a ball is drawn at random and replaced. If it was type i, the ith treatment is assigned. We then wait for a random variable ξ (whose probability distribution depends on i) to be observed. An additional d_{ij} balls are added to the urn of type $j = 1, ..., K$, where $d_{ij}(\xi)$ is some function on the sample space of ξ. The algorithm is repeated through n stages.

Let $Z_n = (Z_{n1}, ..., Z_{nK})$ be the urn composition when the nth patient arrives to be randomized. Then the probability that the patient will be randomized to treatment j is given by $Z_{nj}/|Z_n|$, where $|Z_n| = \sum_{i=1}^{K} Z_{ni}$.

Let $D(\xi) = ((d_{ij})), i, j = 1, ..., K$. First order asymptotics for the generalized Friedman's urn are determined by the generating matrix of the urn, given by $H = E\{D(\xi)\}$. Under certain regularity conditions (H is positive regular and $\Pr\{d_{ij} = 0 \,\forall\, j\} = 0$ for all i),

$$\lim_{n \to \infty} \frac{N_j(n)}{n} = v_j \text{ almost surely}, \tag{10.8}$$

$j = 1, ..., K$, where $v = (v_1, ..., v_K)$ is the normalized (i.e., $\sum_{j=1}^{K} v_j = 1$) *left* eigenvector corresponding to the maximal eigenvalue of H [Athreya and Karlin (1967)].

The generalized Friedman's urn is a natural design for clinical trials of K treatments. Wei (1979) proposed the following simple example of a rule for K treatments. Suppose ξ is a binary outcome, success or failure, and let $d_{ij} = (K-1)\delta_{ij}$ if success on treatment i, and $d_{ij} = (1-\delta_{ij})$ if failure on treatment i, where δ_{ij} is the Kronecker delta. Assuming that ξ is immediately observable after the patient is randomized, we have $|Z_n| = |Z_1| + (K-1)(n-1)$.

10.5.2 The randomized play-the-winner rule

When $K = 2$, and for $\mathbf{Z}_1 = (\alpha, \alpha)$, we have the randomized play-the-winner rule described in Section 10.2.3. This has been the most-studied urn model in the response-adaptive randomization literature. We can explore some of its properties, under a simple population model by letting p_A and p_B be the probabilities of success on treatments A and B, respectively, and $q_A = 1 - p_A$, $q_B = 1 - q_B$. We can write the matrix $\mathbf{H}$ for the $RPW(0, 1)$ rule as

$$\mathbf{H} = \left[\begin{array}{cc} p_A & q_A \\ q_B & p_B \end{array} \right]. \tag{10.9}$$

The maximal eigenvalue of this matrix (since it is stochastic) is 1, and the normalized left eigenvector corresponding to the eigenvalue 1 is $q_B/(q_A + q_B)$. Thus, by (10.8), we obtain

$$\lim_{n \to \infty} \frac{N_A(n)}{n} = \frac{q_B}{q_A + q_B}, \tag{10.10}$$

almost surely, or that

$$\lim_{n \to \infty} \frac{N_A(n)}{N_B(n)} = \frac{q_B}{q_A},$$

almost surely. Consequently, in the limit, the $RPW(0, 1)$ rule allocates according to the relative risk of failure. While this is not optimal in the sense of Section 10.3, this is an intuitively appealing limit.

While finite sample results are intractable for most urn models, the $RPW(\alpha, \beta)$ is simple enough that one can obtain $E(N_A(n))$ and $\mathrm{Var}(N_A(n))$ exactly. A recursion can be developed to determine $E(N_A(n))$ as follows. Again, let $Y_1, ..., Y_n$ be the responses of the n patients, where $Y_j = 1$ if success and 0 if failure. Let $T_1, ..., T_n$ be the treatment assignments, $T_j = 1$ if A and 0 if B and define $\mathcal{F}_n = (Y_1, ..., Y_n, T_1, ..., T_n)$. Note that

$$\begin{aligned} E(T_j Y_j) &= \Pr(T_j = 1, Y_j = 1) \\ &= \Pr(Y_j = 1 | T_j = 1) \Pr(T_j = 1) \\ &= p_A E(T_j). \end{aligned} \tag{10.11}$$

Also we can show that

$$E(Y_j) = p_B + (p_A - p_B)E(T_j) \tag{10.12}$$

(Problem 10.6). The probability of selecting a type A ball is simply the proportion of balls of type A in the urn. After $n - 1$ patients, we have $2\alpha + \beta(n - 1)$ total balls in the urn, and the number of type A balls is the starting number, α, plus the number of successes on A, plus the number of failures on B. Hence, we have

$$E(T_n | \mathcal{F}_{n-1}) = \frac{\alpha + \beta \sum_{i=1}^{n-1} T_i Y_i + \beta \sum_{i=1}^{n-1} (1 - T_i)(1 - Y_i)}{2\alpha + \beta(n - 1)}. \tag{10.13}$$

Taking another expectation, we obtain unconditionally

$$E(T_n) = EE(T_n|\mathcal{F}_{n-1})$$

$$= \frac{\alpha - \beta \sum_{i=1}^{n-1} E(T_i) - \beta \sum_{i=1}^{n-1} E(Y_i) + 2\beta \sum_{i=1}^{n-1} E(Y_i T_i)}{2\alpha + \beta(n-1)},$$

$$= \frac{\alpha + \beta(n-1)q_B}{2\alpha + \beta(n-1)} + \frac{\beta(p_A - q_B)}{2\alpha + \beta(n-1)} \sum_{i=1}^{n-1} E(T_i),$$

using (10.11) and (10.12). Noting that

$$E(N_A(n)) = \sum_{i=1}^{n} E(T_i) = E(T_n) + \sum_{i=1}^{n-1} E(T_i),$$

we can write this as

$$E(N_A(n)) = \frac{\alpha + \beta(n-1)q_B}{2\alpha + \beta(n-1)} + \left(1 + \frac{\beta(p_A - q_B)}{2\alpha + \beta(n-1)}\right) \sum_{i=1}^{n-1} E(T_i).$$

We see that we have a recursion of the form

$$C_n = A_n + B_n C_{n-1}$$

with $A_1 = C_1 = 1/2$. The solution to this recursion is

$$C_n = \sum_{i=1}^{n} A_i \prod_{k=i+1}^{n} B_k.$$

Consequently, we have shown that

$$E(N_A(n)) = \sum_{i=1}^{n} \frac{\alpha + \beta(i-1)q_B}{2\alpha + \beta(i-1)} \prod_{k=i+1}^{n} \left(1 + \frac{\beta(p_A - q_B)}{2\alpha + \beta(k-1)}\right)$$

(Rosenberger and Sriram (1997)). The form of $\mathrm{Var}(N_A(n))$ is more complicated, and is given in Matthews and Rosenberger (1997), requiring at least half a page. They also show that, if $p_A + p_B > 3/2$, the variance depends on the initial urn composition in the limit. This is an undesirable property of urn models for response-adaptive randomization, because the initial urn composition is generally difficult to select in practice, and one would hope that, at least for large samples, that the limiting allocation would be invariant to the starting values. We will discuss the selection of α and β in Chapter 12.

Table 10.3 gives $E(N_A(n)/n)$ and S. D.$(N_A(n)/n)$ for the randomized play the winner rule with $n = 25$. One can see that the variability is quite large for large values of p_A and p_B. Variability is reduced substantially when $\alpha = 5$, but the

Table 10.3 *Exact values of $E(N_A(n)/n)$ (standard deviation) for $n = 25$ for the RPW$(1,1)$ rule and the RPW$(5,1)$ rule (Rosenberger (1999, p. 334), reprinted with permission of Elsevier Science, Inc.).*

p_A	p_B	$\alpha = 1$	$\alpha = 5$
0.1	0.3	0.44 (0.09)	0.46 (0.09)
0.1	0.5	0.38 (0.10)	0.42 (0.09)
0.1	0.7	0.29 (0.10)	0.36 (0.10)
0.1	0.9	0.19 (0.10)	0.31 (0.10)
0.3	0.5	0.43 (0.12)	0.45 (0.10)
0.3	0.7	0.35 (0.13)	0.40 (0.11)
0.3	0.9	0.24 (0.13)	0.34 (0.11)
0.5	0.7	0.41 (0.16)	0.45 (0.13)
0.5	0.9	0.30 (0.17)	0.39 (0.13)
0.7	0.9	0.38 (0.21)	0.44 (0.15)

Table 10.4 *Simulated values of expected allocation proportions, $E(N_A(n)/n)$ (standard deviation), for the RPW$(1,1)$ procedure, 5000 replications (Rosenberger, Stallard, Ivanova, et al., (2001, p. 911), reprinted with permission of International Biometric Society).*

p_A	p_B	n	RPW$(1,1)$
0.1	0.2	526	0.47 (0.02)
0.1	0.3	162	0.44 (0.04)
0.1	0.4	82	0.40 (0.05)
0.4	0.6	254	0.40 (0.05)
0.6	0.9	82	0.29 (0.13)
0.7	0.9	162	0.32 (0.13)
0.8	0.9	526	0.38 (0.12)

adaptive nature of the design is dampened by less extreme allocation to the superior treatment.

Table 10.4 gives the simulated mean allocation proportions (standard deviation) for the RPW$(1, 1)$ rule, which is useful for direct comparison with Table 10.2. It is clear that the rule is more variable than the sequential maximum likelihood procedure for large values of p_A and p_B, but is less variable (but also more conservative) for small values of p_A and p_B.

10.5.3   Ternary urn models

Another class of urn models for clinical trials of two treatment is the *ternary urn*, described by Ivanova and Flournoy (2001). Suppose there are three possible outcomes, A, B, and C. A ball is drawn and replaced, and the appropriate treatment assigned. If the patient's outcome is A at treatment $i, i = 1, ..., K$, a type i ball is added to the urn; if the outcome is B, nothing is done; if the outcome is C, a type i ball is removed. When this model is reduced to only two outcomes, we have three types of urn models. For two outcomes A and B, we have Durham and Yu's (1990) urn, where a ball is added if there is a success and the urn remains unchanged if there is a failure. For outcomes A and C, we have the *birth and death urn* of Ivanova, Rosenberger, Durham, *et al.* (2000), where a ball is added if there is a success and a ball is removed if there is a failure. Finally, if we have outcomes B and C, we have the *drop-the-loser rule* (e.g., Ivanova (2002)), in which a ball is removed if there is a failure, and the urn remains the same if there is a success.

It can be shown that the limiting composition of Durham and Yu's urn will contain only balls representing the best treatment (Durham, Flournoy, and Li (1998), provided there is a single best treatment (*i.e.*, for success probabilities $p_1, ..., p_K$, there exists a unique maximum $p^* \in (p_1, ..., p_K)$). The almost sure limiting allocation is given by

$$\lim_{n\to\infty} \frac{N_j(n)}{n} = 1, \text{ if } p^* = p_j;$$
$$= 0, \text{ otherwise.}$$

The birth and death urn has the following limiting allocation. If $p^* < 1/2$,

$$\lim_{n\to\infty} \frac{N_j(n)}{n} = \frac{\frac{1}{p_j - q_j}}{\sum_{i=1}^{K} \frac{1}{p_i - q_i}}.$$

If $p^* \geq 1/2$, we obtain the same limiting allocation as for Durham and Yu's urn. The limiting allocation of the drop-the-loser rule is given by

$$\lim_{n\to\infty} \frac{N_j(n)}{n} = \frac{\frac{1}{q_j}}{\sum_{i=1}^{K} \frac{1}{q_i}}.$$

For $K = 2$, this is the same limiting allocation as for the randomized play-the-winner rule, given in (10.10).

Urn models that remove balls have a positive probability that certain types of balls will become extinct. In order to eliminate this possibility, Ivanova, Rosenberger, Durham, *et al.* introduced *immigration balls* that replenish the urn according to a Poisson immigration process. This is equivalent to introducing an additional type of ball in the urn, *immigration balls*. An urn contains balls of $K+1$ types, representing K treatments, and ball type $K + 1$ representing immigration balls. If ball types $1, ..., K$ are drawn, the appropriate rule (birth and death or drop-the-loser) is implemented. If an immigration ball is drawn, the ball is returned to the urn along with K additional balls, one of each of types $1, ..., K$.

For $K = 2$, Ivanova and Rosenberger (2001) demonstrate that the drop-the-loser rule, induces less variability when p_A and p_B are large. Since the randomized play-the-winner rule and drop-the loser rule have identical limiting allocations, but the randomized play-the-winner rule is quite variable for large values of p_A and p_B, it seems reasonable to conclude that the drop-the-loser rule is preferable for highly successful treatments in binary response trials, in terms of variability.

Many other urn models have been suggested for response-adaptive randomization. The interested reader is referred to Andersen, Faries, and Tamura (1994), Bandyopadhyay and Biswas (2000), and Bai, Hu, and Shen (2002). These and other urn models are reviewed in Rosenberger (2002).

10.6 TREATMENT EFFECT MAPPINGS

Rosenberger (1993) introduced the idea of a *treatment effect mapping*, in which allocation probabilities are some function of the current treatment effect. Let g : $\Omega \to [0, 1]$, continuous, such that $g(0) = 1/2, g(x) > 1/2$ if $x > 0$, and $g(x) < 1/2$ if $x < 0$. Let $S \in \Omega$ be some measure of the true treatment effect, and let S_j be the observed value of the treatment effect measure after j responses, where $S_j = 0$ if the treatments are equal, $S_j > 0$ if A is better than B, and $S_j < 0$ if A is worse than B. Then we allocate to treatment A with probability

$$E(T_j|\mathcal{F}_{j-1}) = g(S_{j-1}).$$

One would presume that such a procedure would have limiting allocation

$$\lim_{n \to \infty} \frac{N_A(n)}{n} = g(S),$$

but this has not yet been proved formally for general functions g and treatment effect S.

For continuous outcomes, Rosenberger (1993) developed a treatment effect mapping for the linear rank test, using $g(x) = 0.5(1 + x)$, where S is the normalized (centered and scaled) linear rank statistic. Bandyopadhyay and Biswas (2001) used the mapping $g(x) = \Phi(x)$, where Φ is the normal distribution function, and S is the usual two-sample t-test.

For survival outcomes, Rosenberger and Seshaiyer (1997) use the mapping $g(x) = 0.5(1 + x)$ where S is the a centered and scaled logrank test. Yao and Wei (1996) suggest using

$$
\begin{aligned}
g(x) &= 0.5 + xr, \text{ if } xr \in [-0.4, 0.4]; \\
&= 0.1, \text{ if } xr < -0.4; \\
&= 0.9, \text{ if } xr > 0.4,
\end{aligned}
$$

where S is the standardized Gehan Wilcoxon test and r is a constant reflecting the degree to which one wishes to adapt the trial.

The intuitive appeal of the treatment effect mapping approach is that we allocate according to the magnitude of the treatment effect thus far in the trial. While this is an intuitively attractive allocation procedure, the limiting allocation is not optimal, and hence this is a design-driven response-adaptive randomization procedure. Because urn models tend to be relevant for binary and multinomial responses, treatment effect mappings have been proposed for more general outcomes, such as continuous outcomes and survival outcomes. One can also use a covariate-adjusted treatment effect instead of the marginal treatment effect so that the adaptation is based not only on patient responses, but also on their covariates. Covariate-adjusted treatment effect mappings were described in Rosenberger, Vidyashankar, and Agarwal (2001) and Bandyopadhyay and Biswas (2001).

10.7 PROBLEMS

10.1 Use the delta method to derive the expressions for the asymptotic variance of the relative risk and odds ratio measures, given in Table 10.1.

10.2 Show that the optimal allocation for the odds ratio measure using the criterion of Section 10.3 is given by

$$R^* = \sqrt{\frac{p_B \, q_B}{p_A \, q_A}}.$$

10.3 For $p_A = 0.1, 0.5, 0.9$, draw plots superimposing the following allocations across values of p_B:
(i) optimal allocation for the simple difference given in Table 10.1;
(ii) optimal allocation for the relative risk given in Table 10.1;
(iii) optimal allocation for the odds ratio given in Table 10.1;
(iv) limiting allocation for the randomized play-the-winner rule, given in (10.10).
Interpret.

10.4 Show that Neyman allocation assigns more patients to the inferior treatment when $p_A > q_B$.

10.5 a. Show (10.9).
b. Show that the normalized left eigenvector corresponding to the eigenvalue 1 in (10.9) is given by $q_B/(q_A + q_B)$.

10.6 Show (10.12).

10.7 Show that the solution to the recursion

$$C_n = A_n + B_n C_{n-1}$$

with $A_1 = C_1 = 1/2$ is given by

$$C_n = \sum_{i=1}^{n} A_i \prod_{k=i+1}^{n} B_k.$$

10.8 Generate a randomization sequence for $n = 50$ for a binary response trial with $p_A = 0.5$ and $p_B = 0.7$ using the following response-adaptive randomization procedures:

(i) the sequential maximum likelihood procedure targeting $R^* = (p_A/p_B)^{1/2}$;

(ii) RPW (1,1).

10.8 REFERENCES

AGRESTI, A. (1990). *Categorical Data Analysis*. Wiley, New York.

ANDERSON, J., FARIES, D., AND TAMURA, R. N. (1994). A randomized play-the-winner design for multi-armed clinical trials. *Communications and Statistics – Theory and Methods* **23** 309–323.

ANSCOMBE, F. J. (1963). Sequential medical trials. *Journal of the American Statistical Association* **58** 365–384.

ATHREYA, K. B. AND KARLIN, S. (1967). Limit theorems for the split times of branching processes. *Journal of Mathematics and Mechanics* **17** 257–277.

ATHREYA, K. B. AND KARLIN, S. (1968). Embedding of urn schemes into continuous time Markov branching processes and related limit theorems. *Annals of Mathematical Statistics* **39** 1801–1817.

BAI, Z. D., HU, F., AND SHEN, L. (2002). An adaptive design for multi-armed clinical trials. *Journal of Multivariate Analysis*, in press.

BANDYOPADHYAY, U. AND BISWAS, A. (2000). A class of adaptive designs. *Sequential Analysis* **19** 45–62.

BANDYOPADHYAY, U. AND BISWAS, A. (2001). Adaptive designs for normal responses with prognostic factors. *Biometrika* **88** 409–419.

BATHER, J. (1995). Response adaptive allocation and selection bias. In *Adaptive Designs* (FLOURNOY, N. AND ROSENBERGER, W. F., EDS.). Institute of Mathematical Statistics, Hayward, pp. 23–35.

BELLMAN, R. (1956). A problem in the sequential design of experiments. *Sankhya A* **16** 221–229.

BERRY, D. A. (1978). Modified two-armed bandit strategies for certain clinical trials. *Journal of the American Statistical Association* **73** 339–345.

BERRY, D. A. AND EICK, S. G. (1995). Adaptive assignment versus balanced randomization in clinical trials: a decision analysis. *Statistics in Medicine* **14** 231–246.

BERRY, D. A. AND FRISTEDT, B. (1986). *Bandit Problems: Sequential Allocation of Experiments*. Chapman and Hall, London.

CHERNOFF, H. AND ROY, S. N. (1965). A Bayes sequential sampling inspection plan. *Annals of Mathematical Statistics* **36** 1387–1407.

COAD, D. S. (1991). Sequential tests for an unstable response variable. *Biometrika* **78** 113–121.

COLTON, T. (1963). A model for selecting one of two medical treatments. *Journal of the American Statistical Association* **58** 388–400.

CORNFIELD, J., HALPERIN, M., AND GREENHOUSE, S. W. (1969). An adaptive procedure for sequential clinical trials. *Journal of the American Statistical Association* **64** 759–770.

DURHAM, S. D., FLOURNOY, N., AND LI, W. (1998). Sequential designs for maximizing the probability of a favorable response. *Canadian Journal of Statistics* **3** 479–495.

DURHAM, S. D. AND YU, C. F. (1990). Randomized play-the-leader rules for sequential sampling from two populations. *Probability in the Engineering and Information Sciences* **4** 355–367.

EISELE, J. R. (1994). The doubly adaptive biased coin design for sequential clinical trials. *Journal of Statistical Planning and Inference* **38** 249–261.

EISELE, J. R. AND WOODROOFE, M. B. (1995). Central limit theorems for doubly adaptive biased coin designs. *Annals of Statistics* **23** 234–254.

FLOURNOY, N. AND ROSENBERGER, W. F., EDS. (1995). *Adaptive Designs.* Institute of Mathematical Statistics, Hayward.

FLEHINGER, B. J. AND LOUIS, T. A. (1971). Sequential treatment allocation in clinical trials. *Biometrika* **58** 419–426.

GITTINS, J. C. (1989). *Multi-Armed Bandit Allocation Indices.* Wiley, Chichester.

HARDWICK, J. (1995). A modified bandit as an approach to ethical allocation in clinical trials. In *Adaptive Designs* (FLOURNOY, N. AND ROSENBERGER, W. F., EDS.). Institute of Mathematical Statistics, Hayward, pp. 223–237.

HARDWICK, J. AND STOUT, Q. F. (1995). Exact computational analysis for adaptive designs. In *Adaptive Designs* (FLOURNOY, N. AND ROSENBERGER, W. F., EDS.). Institute of Mathematical Statistics, Hayward, pp. 65–87.

HARDWICK, J. AND STOUT, Q. F. (1999). Using path induction for evaluating sequential allocation procedures. *SIAM Journal of Scientific Computing* **21** 67–87.

HAYRE, L. S. (1979). Two-population sequential tests with three hypotheses. *Biometrika* **66** 465–474.

HAYRE, L. S. AND TURNBULL, B. W. (1981). Estimation of the odds ratio in the two-armed bandit problem. *Biometrika* **68** 661–668.

IVANOVA, A. V. (2002). A play-the-winner type urn model with reduced variability. *Metrika*, in press.

IVANOVA, A. AND FLOURNOY, N. (2001). A birth and death urn for ternary outcomes: stochastic processes applied to urn models. In *Probability and Statistical Models with Applications* (CHARALAMBIDES, C. A., KOUTRAS, M. V., AND BALAKRISHNAN, N., EDS.). Chapman and Hall/CRC, Boca Raton, pp. 583–600.

IVANOVA, A. V., ROSENBERGER, W. F., DURHAM, S. D., AND FLOURNOY, N. (2000). A birth and death urn for randomized clinical trials: Asymptotic methods. *Sankhya B* **62** 104–118.

JENNISON, C. AND TURNBULL, B. W. (2000). *Group Sequential Methods with Applications to Clinical Trials.* Chapman and Hall/CRC, Boca Raton.

LACHIN, J. M. (2000). *Biostatistical Methods: The Assessment of Relative Risks.* Wiley, New York.

LOUIS, T. A. (1975). Optimal allocation in sequential tests comparing the means of two Gaussian populations. *Biometrika* **62** 359–369.

MATTHEWS, P. C. AND ROSENBERGER, W. F. (1997). Variance in randomized play-the-winner clinical trials. *Statistics and Probability Letters* **35** 233–240.

MELFI, V. AND PAGE, C. (1995). Variability in adaptive designs for estimation of success probabilities. In *New Developments and Applications in Experimental Design* (FLOURNOY, N., ROSENBERGER, W. F., AND WONG, W. K., EDS.). Institute of Mathematical Statistics, Hayward, pp. 106–114.

MELFI, V. F., PAGE, C., AND GERALDES, M. (2001). An adaptive randomized design with application to estimation. *Canadian Journal of Statistics* **29** 107–116.

ROBBINS, H. (1952). Some aspects of the sequential design of experiments. *Bulletin of the American Mathematical Society* **58** 527–535.

ROBBINS, H. AND SIEGMUND, D. O. (1974). Sequential tests involving two populations. *Journal of the American Statistical Association* **69** 132–139.

ROSENBERGER, W. F. (1993). Asymptotic inference with response-adaptive treatment allocation designs. *Annals of Statistics* **21** 2098–2107.

ROSENBERGER, W. F. (1999). Randomized play-the-winner clinical trials: review and recommendations. *Controlled Clinical Trials* **20** 328–342.

ROSENBERGER, W. F. (2002). Randomized urn models and sequential design. *Sequential Analysis*, in press (with discussion).

ROSENBERGER, W. F. AND SESHAIYER, P. (1997). Adaptive survival trials. *Journal of Biopharmaceutical Statistics* **7** 617–624.

ROSENBERGER, W. F. AND SRIRAM, T. N. (1997). Estimation for an adaptive allocation design. *Journal of Statistical Planning and Inference* **59** 309–319.

ROSENBERGER, W. F., STALLARD, N., IVANOVA, A., HARPER, C. N., AND RICKS, M. L. (2001). Optimal adaptive designs for binary response trials. *Biometrics* **57** 909–913.

ROSENBERGER, W. F., VIDYASHANKAR, A. N., AND AGARWAL, D. K. (2001). Covariate-adjusted response-adaptive designs for binary response. *Journal of Biopharmaceutical Statistics* **11** 227–236.

SIMONS, G. (1989). A random horizon model for sequential clinical trials. *Sequential Analysis* **8** 27–49.

THOMPSON, W. R. (1933). On the likelihood that one unknown probability exceeds another in the view of the evidence of the two samples. *Biometrika* **25** 275–294.

WEI, L. J. (1979). The generalized Pólya's urn design for sequential medical trials. *Annals of Statistics* **7** 291–196.

WEI, L. J. AND DURHAM, S. D. (1978). The randomized play-the-winner rule in medical trials. *Journal of the American Statistical Association* **73** 840–843.

YANG, Y. AND ZHU, D. (2002). Randomized allocation with nonparametric estimation for a multi-armed bandit problem with covariates. *Annals of Statistics* **30** 100-121.

YAO, Q. AND WEI, L. J. (1996). Play the winner for phase II/III clinical trials.

Statistics in Medicine **15** 2413–2423.

ZELEN, M. (1969). Play the winner and the controlled clinical trial. *Journal of the American Statistical Association* **64** 131–146.

11

Inference for Response-Adaptive Randomization

11.1 INTRODUCTION

Inference for response-adaptive randomization is very complicated because both the treatment assignments and responses are correlated. This leads to nonstandard problems and new insights into conditioning. We first examine likelihood-based inference and then randomization-based inference. Most of the work on inference for response-adaptive randomization has focused on urn models. Extension to other response-adaptive designs is certainly feasible, but many of these extensions are open problems. While we mention large sample inference briefly in this chapter, the main results in large sample inference for response-adaptive randomization are proved in Chapter 15.

11.2 POPULATION-BASED INFERENCE

11.2.1 The likelihood

As in Section 7.2, we can use conditioning arguments to derive the likelihood for a response-adaptive randomization. Let $t^{(j)} = (t_1, ..., t_j)$ and $y^{(j)} = (y_1, ..., y_j)$ be the realized treatment assignments and responses from patients $1, ..., j$, respectively. Let θ be the parameter vector of interest. Unlike the restricted randomization case, here $(t_1, ..., t_n)$ depend on θ. However, we have additional data arising from the experiment: the adaptive mechanism. For the urn model, let $z^{(j)} = (z_0, ..., z_{j-1})$ be the history of the urn composition, where z_0 is the initial urn composition and z_i is the urn composition after i stages. Then the likelihood of the data after n patients, denoted $\mathcal{L}_n$, is given by

$$
\begin{aligned}
\mathcal{L}_n &= \mathcal{L}(y^{(n)}, t^{(n)}, z^{(n)}; \theta) \\
&= \mathcal{L}(y_n | y^{(n-1)}, t^{(n)}, z^{(n)}; \theta) \mathcal{L}(t_n | y^{(n-1)}, t^{(n-1)}, z^{(n)}; \theta) \\
&\quad \times \mathcal{L}(z_{n-1} | y^{(n-1)}, t^{(n-1)}, z^{(n-1)}; \theta) \mathcal{L}_{n-1}.
\end{aligned}
\tag{11.1}
$$

Since the responses depend only on the treatment assigned and are independent and identically distributed under a population model, we have

$$
\mathcal{L}(y_n | y^{(n-1)}, t^{(n)}, z^{(n)}; \theta) = \mathcal{L}(y_n | t_n; \theta).
\tag{11.2}
$$

Now the treatment assignments will depend only on the current urn composition at the time of assignment. This means that

$$
\mathcal{L}(t_n | y^{(n-1)}, t^{(n-1)}, z^{(n)}; \theta) = \mathcal{L}(t_n | z_{n-1}).
\tag{11.3}
$$

Noting that the urn composition at stage $n - 1$ is completely determined by the urn composition at stage $n - 2$ and the treatment assignment and response of the $n - 1$th patient, we see that

$$
\mathcal{L}(z_{n-1} | y^{(n-1)}, t^{(n-1)}, z^{(n-1)}; \theta) = 1.
\tag{11.4}
$$

Combining (11.1)–(11.4), we obtain

$$
\begin{aligned}
\mathcal{L}_n &= \mathcal{L}(y_n | t_n; \theta) \mathcal{L}(t_n | z_{n-1}) \mathcal{L}_{n-1} \\
&= \prod_{i=1}^{n} \mathcal{L}(y_i | t_i; \theta) \mathcal{L}(t_i | z_{i-1}).
\end{aligned}
$$

Since $\mathcal{L}(t_i | z_{i-1})$ is independent of θ, we have

$$
\mathcal{L}_n \propto \prod_{i=1}^{n} \mathcal{L}(y_i | t_i; \theta)
\tag{11.5}
$$

(Rosenberger, Flournoy, and Durham, 1997).

Note that (11.5) is identical to the likelihood from restricted randomization, in (7.5). However, this only means that the likelihoods look the same. The distribution of the sufficient statistics is quite different in response-adaptive randomization than in restricted randomization, as we shall see.

For the case where there are K treatments, let $\delta_{ji} = 1$ if $t_i = j, j = 1, ..., K$, and 0 otherwise. When we have binary responses and $\Pr(Y_i = 1 | T_i = j) = p_j$, we can write the likelihood as

$$
\begin{aligned}
\mathcal{L}_n &= \prod_{i=1}^{n} \prod_{j=1}^{K} [\mathcal{L}(y_i | t_i; \theta)]^{\delta_{ji}} \\
&= \prod_{i=1}^{n} \prod_{j=1}^{K} p_j^{y_i \delta_{ji}} (1 - p_j)^{(1 - y_i)\delta_{ji}} \\
&= \prod_{j=1}^{K} p_j^{s_j(n)} (1 - p_j)^{n_j(n) - s_j(n)},
\end{aligned}
\tag{11.6}
$$

where $s_j(n) = \sum_{i=1}^{n} y_i \delta_{ji}$ and $n_j(n) = \sum_{i=1}^{n} \delta_{ji}$. Let $S_j(n)$ and $N_j(n)$ be the random analogs of $s_j(n)$ and $n_j(n)$, respectively. Note that the maximum likelihood estimator of p_j is $\hat{p}_j = S_j(n)/N_j(n)$. Under certain regularity conditions, the maximum likelihood estimators are consistent and asymptotically normal. The details are given in Chapter 15.

11.2.2 Sufficiency

We can determine the sufficient statistics for θ from the likelihood. Here is where we must carefully distinguish between restricted and response-adaptive randomization. For restricted designs, $N_j(n)$ does not depend on p_j, and hence is an ancillary statistic. It follows that $S_j(n)$ is a complete sufficient statistic for p_j. By Basu's Theorem (Lehmann (1983, p. 46)), this implies that $S_j(n)$ and $N_j(n)$ are independent.

In contrast, for response-adaptive randomization, $N_j(n)$ does carry information about p_j, and therefore is not ancillary. In fact, the statistics $(S_1(n),...,S_K(n), N_1(n),...,N_{K-1}(n))$ are jointly sufficient for $(p_1,...,p_K)$. This brings up the interesting dilemma that if we condition on $N_j(n)$ when we do inference, we lose extensive information. Thus response-adaptive randomization requires *unconditional* tests.

11.2.3 Bias of the maximum likelihood estimators

Because of the dependence structure induced by response-adaptive randomization, the maximum likelihood estimators, although they are typically consistent, are biased. Coad and Ivanova (2001) derive the bias factor as follows. Let $S_A(n)$, $S_B(n)$, $N_A(n)$, $N_B(n)$ be the number of success on A, successes on B, number of patients on A, patients on B, respectively. Let $\mathcal{P}_{p_A,p_B}$ be the probability measure on the sequences of treatment responses determined by p_A and p_B and E_{p_A,p_B} be the expectation with respect to that measure. Then $d\mathcal{P}_{p_A,p_B}$ is given by

$$
d\mathcal{P}_{p_A,p_B} = \frac{N_A(n), N_B(n),}{S_A(n), S_B(n), (N_A(n) - S_A(n)), (N_B(n) - S_B(n)),}
$$
$$
\times p_A^{S_A(n)}(1 - p_A)^{N_A(n) - S_A(n)} p_B^{S_B(n)}(1 - p_B)^{N_B(n) - S_B(n)}
$$

and the first derivative is given by

$$
\frac{\partial}{\partial p_i} d\mathcal{P}_{p_A,p_B} = \frac{1}{p_i(1 - p_i)}(S_i(n) - p_i N_i(n)) d\mathcal{P}_{p_A,p_B}, i = A, B.
$$

We can write

$$
E_{p_A,p_B}\left(\frac{1}{N_i(n)}\right) = \int \frac{1}{N_i(n)} d\mathcal{P}_{p_A,p_B}, i = A, B.
$$

Assuming that E_{p_A,p_B} is continuous in p_i, we may differentiate under the integral sign to obtain

$$
\frac{\partial}{\partial p_i} E_{p_A,p_B}\left(\frac{1}{N_i(n)}\right) = \frac{1}{p_i(1-p_i)} \int \left(\frac{1}{N_i(n)}\right)(S_i(n) - p_i N_i(n)) dP_{p_A,p_B}
$$

$$
= \frac{1}{p_i(1-p_i)} E_{p_A,p_B}(\hat{p}_i - p_i). \tag{11.7}
$$

Using (11.7), we can write the bias as

$$
E_{p_A,p_B}(\hat{p}_i - p_i) = p_i(1-p_i)\frac{\partial}{\partial p_i} E_{p_A,p_B}\left(\frac{1}{N_i(n)}\right). \tag{11.8}
$$

Clearly this is zero if we do not have response-adaptive randomization.

The delta method will allow us to obtain a suitable approximation to (11.8). We can write

$$
E_{p_A,p_B}\left(\frac{1}{N_i(n)}\right) \simeq \frac{1}{E_{p_A,p_B}(N_i(n))} + \frac{\mathrm{Var}_{p_A,p_B}(N_i(n))}{[E_{p_A,p_B}(N_i(n))]^3}, i = A, B, \tag{11.9}
$$

(Problem 11.1). For specific response-adaptive randomization procedures, if we can compute the variance of $N_i(n)$, we can obtain an approximate bias correction using (11.8) and (11.9).

As an example, consider the randomized play-the-winner rule with two treatments, where $q_A = 1 - p_A$ and $q_B = 1 - p_B$. It is known that

$$
\frac{E(N_A(n))}{n} \to \frac{q_B}{q_A + q_B}
$$

and that, when $p_A + p_B < 3/2$,

$$
\frac{\mathrm{Var}(N_A(n))}{n} \to \frac{q_A q_B(1 + 2p_A + 2p_B)}{(q_A + q_B)^2(3 - 2p_A - 2p_B)}
$$

(Matthews and Rosenberger, 1997). Then we can obtain the following approximation for the bias:

$$
E_{p_A,p_B}(\hat{p}_A - p_A) \simeq p_A q_A \frac{\partial}{\partial p_A}\left[\frac{q_A + q_B}{nq_B} + \frac{q_A(q_A + q_B)(1 + 2p_A + 2p_B)}{n^2 q_B^2(3 - 2p_A - 2p_B)}\right].
$$

If we ignore the term of order $O(n^{-2})$, we obtain

$$
E_{p_A,p_B}(\hat{p}_A - p_A) = -\frac{p_A q_A}{nq_B} + o(n^{-1}).
$$

This correction is reasonably accurate for small sample sizes. Similar bias corrections are given for other urn designs in Coad and Ivanova (2001).

11.2.4   Confidence interval procedures

Confidence interval procedures have been proposed for response-adaptive randomization. These include exact binomial confidence intervals for the randomized play-the-winner rule and a bootstrap procedure for a general response-adaptive randomization procedure of K treatments with binary responses. Coad and Woodroofe (1997) and Coad and Govindarajulu (2000) construct confidence intervals following sequential adaptive designs for censored survival data and binary responses, respectively.

We outline the basic procedure for the computation of exact confidence intervals. Wei, Smythe, Lin, *et al.* (1990) derived exact binomial confidence intervals for difference measures of p_A and p_B, such as the simple difference $\Delta = p_A - p_B$ following a clinical trial using the randomized play-the-winner rule. The exact distribution will depend on Δ and a nuisance parameter, p_B. One popular approach for dealing with a nuisance parameter is to condition on a sufficient statistic. Wei, Smythe, Lin, *et al.* chose to maximize over the possible values of p_B. Let S_A, S_B, N_A, N_B be the number of successes on treatments A and B and the numbers of patients on A and B, respectively, with realizations s_A, s_B, n_A, n_B, and let $\hat{\Delta} = s_A/n_A - s_B/n_B$.

Then if $n_A, n_B > 0$, the exact unconditional confidence interval $(\underline{\Delta}, \overline{\Delta})$ can be computed according to the formulas

$$\underline{\Delta} = \inf_{-1 \leq \Delta \leq 1} \{\Delta : [\max_{p_B} \Pr(S_A/N_A - S_B/N_B \geq \hat{\Delta}, N_A > 0, N_B > 0)] \geq \alpha_1\},$$

$$\overline{\Delta} = \sup_{-1 \leq \Delta \leq 1} \{\Delta : [\max_{p_B} \Pr(S_A/N_A - S_B/N_B \geq \hat{\Delta}, N_A > 0, N_B > 0)] \geq \alpha_2\},$$

for fixed constants α_1 and α_2. These confidence intervals can be computationally intensive, and rely on the networking algorithm approach. Wei, Smythe, Lin, *et al.* compare their unconditional confidence interval to the conditional confidence interval and found that the unconditional intervals tend to be shorter and more efficient than the conditional counterparts for the randomized play-the-winner rule.

Rosenberger and Hu (1999) derived bootstrap confidence intervals following a general response-adaptive randomization procedure of K treatments, using a simple rank ordering. The algorithm is as follows:

1. Obtain the observed data, $\hat{p} = (\hat{p}_1, ..., \hat{p}_K)$ and $N = (N_1, ..., N_K)$, the vector of observed success proportions and sample sizes.
2. Simulate the adaptive allocation rule B times, using $\hat{p}$ as the underlying response probabilities, obtaining B sequences of treatment assignments and responses.
3. Compute $\hat{p}_1^*, ..., \hat{p}_B^*$ and $N_1^*, ..., N_B^*$ from the simulations. These are the bootstrap estimates of the response probabilities and sample sizes.
4. Order $\hat{p}_i^{*1}, ..., \hat{p}_i^{*B}$, for $i = 1, ..., K$ as $\hat{p}_i^{*(1)}, ..., \hat{p}_i^{*(B)}$.

The simplest $100(1 - \alpha)\%$ bootstrap confidence interval approximation for p_i is given by

$$\left(\hat{p}_i^{*(B\alpha/2)}, \hat{p}_i^{*(B(1-\alpha)/2)}\right).$$

Rosenberger and Hu show that this simple confidence interval provides near perfect coverage. The same techniques can be used for measures of treatment differences, and can incorporate delayed response and staggered entry in the simulation phase 2.

11.3 POWER

Response-adaptive randomization induces additional correlation among the responses, and this leads to an increase in the variance of the test statistic. This increased variance contributes to a decrease in power for standard tests based on a population model. In general, for clinical trials of two treatments, power of the test will be intimately linked to $\mathrm{Var}(N_A(n)/n)$. One can see this quite readily when examining the noncentrality parameter for the test of the simple difference in binary response trials (Hu and Rosenberger, 2002).

Suppose we have a fixed target proportion ρ, for instance, ρ could be based on some optimization criterion, or the limiting allocation of an urn design, as discussed in Chapter 10. For this case, we can calculate the noncentrality parameter for the Z-test as follows:

$$\frac{(p_A - p_B)^2}{p_A q_A / N_A(n) + p_B q_B / N_B(n)},$$

which can be rewritten as

$$\frac{(p_A - p_B)^2}{p_A q_A / [n\rho + n(N_A(n)/n - \rho)] + p_B q_B / [n(1 - \rho) - n(N_A(n)/n - \rho)]}.$$

Now define a function

$$g(x) = \frac{(p_A - p_B)^2}{p_A q_A / (\rho + x) + p_B q_B / ((1 - \rho) - x)}.$$

We have the following expansion:

$$g(x) = g(0) + g'(0)x + g''(0)x^2/2 + o(x^2).$$

After some calculation, we obtain

$$g'(0) = (p_A - p_B)^2 \frac{(p_A q_A (1 - \rho)^2 - p_B q_B \rho^2)}{(p_A q_A (1 - \rho) + p_B q_B \rho)^2}$$

and

$$g''(0) = -2(p_A - p_B)^2 \frac{p_A q_A p_B q_B}{((1 - \rho)\rho)^3}.$$

Thus, we have that the noncentrality parameter of the test is given by

$$
n \times \left\{ \frac{(p_A - p_B)^2}{p_A q_A / \rho + p_B q_B / (1 - \rho)} \right.
$$

$$
+ (p_A - p_B)^2 \frac{(p_A q_A (1 - \rho)^2 - p_B q_B \rho^2)}{(p_A q_A (1 - \rho) + p_B q_B \rho)^2} (N_A(n)/n - \rho)
$$

$$
- (p_A - p_B)^2 \frac{p_A q_A p_B q_B}{((1 - \rho)\rho)^3} (N_A(n)/n - \rho)^2
$$

$$
\left. + o((N_A(n)/n - \rho)^2) \right\}. \tag{11.10}
$$

The first term in (11.10) is determined by ρ, which represents the noncentrality parameter for the fixed design. The second term in (11.10) represents the bias of the randomized design from the target proportion. With the design shifting to a different side from the target proportion ρ, the noncentrality parameter will increase or decrease according the coefficient

$$
(p_A - p_B)^2 \frac{(p_A q_A (1 - \rho)^2 - p_B q_B \rho^2)}{(p_A q_A (1 - \rho) + p_B q_B \rho)^2}.
$$

To control the power, it may be desired to have this coefficient be 0. It is interesting to see that this coefficient equals 0 if and only if $p_A q_A (1 - \rho)^2 - p_B q_B \rho^2 = 0$, that is

$$
\rho = \frac{\sqrt{p_A q_A}}{\sqrt{p_A q_A} + \sqrt{p_B q_B}},
$$

i.e., Neyman allocation!

For response-adaptive randomization procedures, we can consider the expectation of the noncentrality parameter. If $E(N_A(n)/n - \rho) = 0$, at least to order $o(1/n)$, then the average power lost of the response-adaptive randomization procedure is then a function of

$$
-(p_A - p_B)^2 \frac{p_A q_A p_B q_B}{((1 - \rho)\rho)^3} E(N_A(n)/n - \rho)^2,
$$

which fully represents the variability of the design. So we have now have the precise link between power and the variation of the design. Thus we can use the variance of $N_A(n)/n$ to compare response-adaptive randomization procedures with same allocation limit or the variance and bias if they do not have the same limiting allocation.

Unfortunately, finding the asymptotic variance of $N_A(n)/n$ is a difficult task for most response-adaptive randomization procedures. Matthews and Rosenberger (1997) present an expression for the randomized play-the-winner rule, and Bai and Hu (1999) extend this to all generalized Friedman's urns, but their takes more than a full page to describe. Hu and Zhang (2002) have derived the asymptotic variance for the doubly-adaptive biased coin design when targeting $\rho = q_B/(q_A + q_B)$, the limiting allocation for the randomized play-the-winner rule and conclude that

Table 11.1 *Simulated power and expected number of failures for the standard Z-test of two proportions under the sequential maximum likelihood procedure targeting (10.4) (A), the sequential maximum likelihood procedure targeting Neyman allocation (N), the randomized play-the-winner allocation (R), and equal allocation (E), 5000 replications (Rosenberger, Stallard, Ivanova, et al. (2001, p. 912), reprinted with permission of International Biometric Society).*

			Power				Expected Failures			
p_A	p_B	n	A	N	R	E	A	N	R	E
0.1	0.2	526	0.89	0.89	0.90	0.90	443.0	444.0	445.6	447.1
0.1	0.3	162	0.89	0.90	0.89	0.90	126.2	127.0	127.7	129.6
0.1	0.4	82	0.89	0.90	0.89	0.90	58.5	59.4	59.0	61.5
0.4	0.6	254	0.89	0.89	0.89	0.89	124.4	126.9	121.9	127.0
0.6	0.9	82	0.90	0.90	0.86	0.90	19.3	22.4	15.3	20.5
0.7	0.9	162	0.91	0.90	0.87	0.90	31.5	35.2	26.6	32.4
0.8	0.9	526	0.90	0.90	0.88	0.90	78.3	82.5	72.6	78.9

the doubly-adaptive biased coin design is always less variable than the randomized play-the-winner rule. Specifically how this influences power has not yet been investigated.

A few researchers have explored power of response-adaptive randomization procedures using simulation. Rosenberger, Stallard, Ivanova, *et al.* (2001) explore the differences in power and expected number of treatment failures for the randomized play-the-winner rule, the sequential maximum likelihood procedure targeting $\rho = R^*$ in (10.4), the sequential maximum likelihood procedure targeting Neyman allocation, and equal allocation for the simple difference of proportions test. Results are given in Table 11.1. One can see that for alternatives where p_A and p_B are small (less than 0.5), the sequential maximum likelihood procedure targeting $(p_A/p_B)^{1/2}$ is the best, with similar power and fewer treatment failures. It should be noted that sequential Neyman allocation is almost as good, with differences of only about one treatment failure. As p_A and p_B get larger, there is little difference between the sequential maximum likelihood procedure and equal allocation, and Neyman allocation results in too many treatment failures. While the randomized play-the-winner rule results in fewer failures, it also is highly variable, and results in power losses of 2-4 percent. When one increases the sample size to equate the power of equal allocation and the randomized play-the-winner rule, the expected failures are worse than for equal allocation.

So for small success probabilities, the sequential maximum likelihood procedure targeting the optimal allocation is the desired randomization procedure. However, it does not result in much of an advantage over equal allocation for highly successful treatment. Ivanova and Rosenberger (2001) show that the drop-the-loser rule results in a very powerful test based on the asymptotic distribution of the odds ratio, when p_A and p_B are large.

For the treatment effect mapping procedure using the logrank test under staggered entry and censoring, Rosenberger and Seshaiyer (1997) found that the power of the test was always within 1 percent of that for equal allocation, and reductions in expected treatment failures was considerable. Similar favorable results were found by Yao and Wei (1996) using a treatment effect mapping with the Gehan Wilcoxon test and by Hallstrom, Brooks, and Peckova (1996), using other types of allocation procedures with the logrank test. However, for a mapping of the difference of normal means, Bandyopadhyay and Biswas (2001) found significant losses in power. In clinical trials of $K > 2$ treatments, Ivanova and Rosenberger (2000) found that the birth and death urn with immigration required increased numbers of patients to have the same power as equal allocation, but the expected number of treatment failures is still less using the urn allocation.

Power for sequential tests in a clinical trial using a sequential monitoring scheme with randomized play-the-winner randomization is explored by simulation in Coad and Rosenberger (1999) and by computing exact distributions using a networking algorithm in Stallard and Rosenberger (2002). While the former paper found a slight advantage for the randomized play-the-winner rule over equal allocation for a sequential test, the latter paper found the randomized play-the-winner rule does not improve expected treatment successes when used in conjunction with a particular sequential monitoring scheme.

In general, the properties of response-adaptive randomization in conjunction with sequential tests based on an early stopping boundary have been sorely lacking in the literature. This is a fundamental area of research that must be tackled before response-adaptive randomization can enjoy popular use in sequentially-monitored clinical trials.

11.4 RANDOMIZATION-BASED INFERENCE

As with restricted randomization procedures, randomization-based inference can be performed following a response-adaptive randomization procedure using the family of linear rank tests. These tests are completely nonparametric, and depend only on the way the n patients were randomized for fixed values of the patient scores. When $K = 2$, we are interested in the linear rank test given by

$$ W = \sum_{j=1}^{n} (a_{jn} - \bar{a}_n) T_j, \tag{11.11} $$

where the a_{jn}'s are fixed constants and $T_j = 1$ if treatment A was assigned and 0 if treatment B. In the case of binary response, we can let a_{jn} be 1 if success and 0 if failure. In this case, letting $S_A(n)$ and $S_B(n)$ be the number of successes on A and

Table 11.2 Unconditional reference set for computation of the linear rank test following $RPW(1,1)$ randomization.

Sequence (l)	$\Pr(L = l)$	S_l
$AAAA$	1/15	0.0
$AAAB$	1/10	-0.5
$AABA$	1/10	0.5
$AABB$	1/15	0.0
$ABAA$	3/40	0.5
$ABAB$	1/20	0.0
$ABBA$	1/30	1.0
$ABBB$	1/120	0.5
$BAAA$	1/120	-0.5
$BAAB$	1/30	-1.0
$BABA$	1/20	0.0
$BABB$	3/40	-0.5
$BBAA$	1/15	0.0
$BBAB$	1/10	-0.5
$BBBA$	1/10	0.5
$BBBB$	1/15	0.0

B, respectively, a little algebra shows that (11.11) is equivalent to

$$W = \frac{N_A(n)N_B(n)}{n} \left(\frac{S_A(n)}{N_A(n)} - \frac{S_B(n)}{N_B(n)} \right). \tag{11.12}$$

Table 11.2 shows the computation of the exact test under $RPW(1,1)$ randomization for $n = 4$ when the patient's responses were $a_{jn} = \{1, 0, 0, 1\}$ and the observed allocation was $T_j = \{A, A, B, B\}$. Then the observed test statistic is $W = 0$. The unconditional p-value is computed by summing the probabilities of each sequence where $S_l \geq 0$. This yields $p_u = 0.6833$.

Wei (1988) proposed a version of this test with uncentered scores, given by

$$W = \sum_{j=1}^{n} a_{jn} T_j = S_A(n). \tag{11.13}$$

This test generated much controversy, which was recorded in the paper by Begg (1990) with ensuing discussion. The test in (11.12) certainly lends itself to a more straightforward interpretation in terms of the observed treatment difference.

As described in Section 7.8, computational algorithms can be developed to compute the exact distribution of permutation tests if samples are not too large. Hardwick and Stout (1998) give the general approach for developing software to find exact distributions with adaptive designs. They have usually been successful using parallel processing for samples up to $n = 200$. The approach for urn models is described in Ivanova and Rosenberger (2000) and was initially proposed by Wei (1988). The techniques are similar as for restricted randomization, but one must track not only the numbers of patients assigned to each treatment, but also the number of successes.

Let $N_n = (N_1(n), ..., N_K(n))$ be the number of patients assigned after n draws for treatments $1, ..., K$, and $S_n = (S_1(n), ..., S_K(n))$ be the number of successes on treatments $1, ..., K$. Consider a network at the $(n + 1)$th stage. At stage n, it is a set of nodes of the form $(N_n, S_n, P(N_n, S_n))$, where $P(N_n, S_n)$ is the probability of a realization of (N_n, S_n). Let Ω_n be the set of all triples $(N_n, S_n, P(N_n, S_n))$ with distinct (N_n, S_n). The set Ω_{n+1} can be obtained recursively from Ω_n. Given (N_n, S_n), one can then determine the number of balls of each type in the urn after n draws, and hence the respective probability of being assigned to treatments $1, ..., K$. Starting with $\Omega_0 = (0, 0, 1)$, we can obtain Ω_{n+1} moving recursively through draws $1, ..., n$. Records with the same (N_i, S_i) are merged into one when moving from the ith stage to the $(i + 1)$th. In this way, one can obtain the exact distribution of N_n and S_n. From this, we can determine the exact distribution of W in (11.12).

The large sample distribution of the linear rank test has been derived for the randomized play-the-winner rule and one other response-adaptive randomization procedure. These are discussed in Chapter 15. In general, there is very little literature related to randomization-based inference for response-adaptive randomization.

11.5 PROBLEMS

11.1 Use the delta method to show (11.9).

11.2 Read Wei (1988) and Begg (1990), along with the ensuing discussion. Write a short paper summarizing the various methods discussed and points for and against each (from Begg and the discussants). What are your views on inference following a response-adaptive randomization procedure?

11.6 REFERENCES

BAI, Z. D. AND HU, F. (1999). Asymptotic theorems for urn models with nonhomogeneous generating matrices. *Stochastic Processes and Their Applications* **80** 87–101.

BANDYOPADHYAY, A. AND BISWAS, A. (2001). Adaptive designs for normal responses with prognostic factors. *Biometrika* **88** 409–419.

BEGG, C. B. (1990). On inferences from Wei's biased coin design for clinical trials. *Biometrika* **77** 467–484 (with discussion).

COAD, D. S. AND GOVINDARAJULU, Z. (2000). Corrected confidence intervals following a sequential adaptive trial with binary response. *Journal of Statistical Planning and Inference* **91** 53–64.

COAD, D. S. AND IVANOVA, A. (2001). Bias calculations for adaptive urn designs. *Sequential Analysis* **20** 229–239.

COAD, D. S. AND ROSENBERGER, W. F. (1999). A comparison of the randomised play-the-winner rule and the triangular test for clinical trials with binary responses. *Statistics in Medicine* **18** 761–769.

COAD, D. S. AND WOODROOFE, M. B. (1997). Approximate confidence intervals after a sequential clinical trial comparing two exponential survival curves with censoring. *Journal of Statistical Planning and Inference* **63** 79–96.

HALLSTROM, A., BROOKS, M. M., AND PECKOVA, M. (1996). Logrank, play the winner, power and ethics. *Statistics in Medicine* **15** 2135–2142.

HARDWICK, J. AND STOUT, Q. F. (1998). Flexible algorithms for creating and analyzing adaptive sampling procedures. In *New Developments and Applications in Experimental Design* (FLOURNOY, N., ROSENBERGER, W. F., AND WONG, W. K., EDS.) Institute of Mathematical Statistics, Hayward, pp. 91–105.

HU, F. AND ROSENBERGER, W. F. (2002). Power and response-adaptive designs. In preparation.

HU, F. AND ZHANG, L-X. (2002). Asymptotic propeties of doubly adaptive biased coin designs in multi-treatment clinical trials. Submitted.

IVANOVA, A. AND ROSENBERGER, W. F. (2000). A comparison of urn designs for randomized clinical trials of $K > 2$ treatments. *Journal of Biopharmaceutical Statistics* **10** 93–107.

IVANOVA, A. AND ROSENBERGER, W. F. (2001). Adaptive designs for highly successful treatments. *Drug Information Journal* **35** 1087–1093.

LEHMANN, E. L. (1983). *The Theory of Point Estimation.* Wiley, New York.

MATTHEWS, P. C. AND ROSENBERGER, W. F. (1997). Variance in randomized play-the-winner clinical trials. *Statistics and Probability Letters* **35** 223–240.

ROSENBERGER, W. F. AND HU, F. (1999). Bootstrap methods for adaptive designs. *Statistics in Medicine* **18** 1757–1767.

ROSENBERGER, W. F. AND SESHAIYER, P. (1997). Adaptive survival trials. *Journal of Biopharmaceutical Statistics* **7** 617–624.

ROSENBERGER, W. F., STALLARD, N., IVANOVA, A., HARPER, C. N., AND RICKS, M. L. (2001). Optimal adaptive designs for binary response trials. *Biometrics* **57** 909–913.

STALLARD, N. AND ROSENBERGER, W. F. (2002). Exact group-sequential designs for clinical trials with randomised play-the-winner allocation. *Statistics in Medicine* **21** 467–480.

WEI, L. J. (1988). Exact two-sample permutation tests based on the randomized play-the-winner rule. *Biometrika* **75** 603–606.

WEI, L. J., SMYTHE, R. T., LIN, D. Y., AND PARK, T. S. (1990). Statistical inference with data-dependent treatment allocation rules. *Journal of the American Statistical Association* **85** 156–162.

YAO, Q. AND WEI, L. J. (1996). Play the winner for phase II/III clinical trials. *Statistics in Medicine* **15** 2413–2423.

12

Response-Adaptive Randomization in Practice

12.1 BASIC ASSUMPTIONS

In this chapter we explore practical considerations in the use of response-adaptive randomization. It should be clear from Chapters 10 and 11 that there are three basic assumptions underlying the use of these types of designs.

First, one must assume that it is feasible to identify the "better" treatment with high probability. This, in turn, will depend on the target sample size for the trial and the treatment effect anticipated. Usually the designed treatment effect is modest, since studies are designed to detect the minimal clinically relevant difference in treatments. The smaller the designed effect, the larger the sample size needed to provide a high probability that the better of the two treatments is so identified.

Second, we must assume that the "better" treatment is not associated with any potential severe toxicity, short or long-term. Otherwise the design will be assigning the majority of patients to an unsafe therapy. In fact, some have suggested the importance of at least beginning the trial with equal allocation until some experience is gained that the treatments are safe, before beginning adaptive randomization.

Third, some patient data on the primary outcome of the trial must be accrued prior to randomizing most of the patients. This immediately eliminates long-term survival trials with limited recruitment and a follow-up extending years. In many of those trials, outcome data become evident only after the recruitment phase has ended. While long-term survival trials represent a large portion of major multi-center clinical trials, there are many shorter duration trials in which the recruitment period can be extended to provide data for the adaptation of future allocation probabilities.

These criteria, particularly the second, would tend to preclude use of response-adaptive randomization in most phase II trials of new drugs where the safety of the agent has yet to be established. However, response-adaptive randomization could be ideal in phase II studies of an established safe agent in a new patient population. Likewise, this would preclude the use of response-adaptive randomization in phase III trials of agents in which animal toxicology or phase II studies have raised the possibility of short- or long-term adverse effects. However, response-adaptive randomization could be ideal for studies of competing agents for a given indication, all of which were previously documented to be safe, or for a phase III compound that is a member of a family of drugs, the safety of which has already been established.

While most of the models examined in Chapter 10 assume that patient responses are ascertainable immediately before the next patient is randomized, that assumption is used only to simplify the probabilistic properties of the response-adaptive randomization rules. In practice, one can "adapt" at certain fixed points in the trial using grouped data already accrued, or one can factor in a delayed response by just using the data available. In the latter setting, one would update the urn (for urn models) or update the maximum likelihood estimators (for sequential maximum likelihood procedures) as each patient responds. Simulation studies have shown that (at least for urn models), whereas the allocation probabilities are not as extreme as for immediate response trials, response-adaptive randomization with delayed response still reduces the expected number of failures and puts more patients on the better treatment when there is delayed response (Rosenberger, 1999).

While certainly a minority of clinical trials are performed with a primary outcome that is ascertainable immediately, a good number of such trials are conducted. Often these are clinical trials of surgical interventions or other medical procedures with an easily ascertainable "success" or "failure" outcome, which is known before the next patient undergoes the procedure. One example is the prevention of hypotension associated with spinal anesthesia for Cesarean section. Rout, Rocke, Levin, *et al.* (1993) describe such a trial of crystalloid preload versus placebo, using Zelen's play-the-winner rule (Section 10.2.1) to allocate treatments.

12.2 BIAS, MASKING, AND CONSENT

Because response-adaptive randomization procedures are randomized, they enjoy many of the same benefits of other randomization procedures in terms of mitigation of bias. However, there are several ways that bias can enter a trial using response-adaptive randomization procedures.

As with any randomization procedure, the clinical trial should, whenever possible, be double-masked. The current allocation probabilities should be kept strictly confidential by the statistician responsible for randomization, as knowledge of the allocation probabilities is tantamount to knowledge of the current treatment effect. Even in unmasked studies, response-adaptive randomization procedures offer some protection from selection bias provided that the responses of previously entered patients are masked. If the responses are unmasked, and their corresponding treatment assignments also unmasked, then one would expect that such designs afford less protection against selection bias than restricted randomization procedures, if the shift in probability of assignment to A away from 0.5 is larger for response-adaptive randomization when one treatment is superior than for restricted randomization when there is a treatment imbalance.

These procedures also provide some protection against accidental bias, provided that one assumes that all subjects arise at random from an underlying homogeneous population, such that the probability of the covariates, and also patient responses, are identical over time. For example, consider the following simple scenario. Assume that a simple two-stage adaptive procedure is employed with the same number of subjects recruited in the first and second stages. In the first stage, the probability of assignment to A is 0.5 and the probability of recruiting a male subject is 0.5. Then at the second stage, based on the finding of more beneficial response with A during the first stage, the probability of assignment to A is modified to 0.8. Now also assume that by chance or due to a change in recruitment strategies the probability of recruiting a male subject during the second stage is 0.7. Thus, during the second stage, it is more likely that a patient will be male than female, and more likely that the patient will be assigned to A rather than B. This will result in a covariate imbalance, in which 62.3 percent of those assigned to A will be male, versus 55.7 percent of those assigned to B. If the probability of treatment response differed greatly between males and females, this would also introduce a bias into the results of the study. One could also evaluate by simulation the susceptibility of response-adaptive randomization to the trend of a covariate, qualitative or quantitative, over time. However, this simple example, with an extreme shift in the assignment probabilities, and an extreme shift in the covariate distribution, still results in a degree of imbalance that would be readily adjusted for in a post-hoc stratification or a regression adjustment, as described in Chapter 8.

Rosenberger and Lachin (1993) suggest that consent forms should state that participants will receive one of two treatments and the probability of treatment assignments will depend on the relative merits of the two treatments based on responses of previously treated volunteers. Such a statement should make the trial more attractive to participants than simply telling them that they are equally likely to receive either treatment. It should also be made clear that the treatment performing better thus far may not, in fact, be the better treatment overall, because the study has not been completed and there are not enough patients currently to make that evaluation.

Informing the patient of the nature of the response-adaptive randomization in this way may lead to a different kind of bias, coined *accrual bias* by Rosenberger (1996), in which volunteers may wish to be recruited later in the trial so as to benefit from the full impact of previous outcomes, and thereby have a better chance to receive the better treatment. Rosenberger (1999) recommends that patients be masked to their sequence number in the trial to prevent accrual bias; whether such masking is acceptable to patients and physicians has not been investigated. Accrual bias is irrelevant in trials dealing with emergency therapies, such as emergency surgical procedures.

12.3   LOGISTICAL ISSUES

There are two main differences between the implementation of response-adaptive randomization and the implementation of other randomization procedures. These differences become magnified as the complexity of the trial increases, particularly in the multi-center situation. First, as pointed out by Faries, Tamura, and Andersen (1995), response-adaptive randomization requires much more communication among the sponsor or coordinating center and the investigators in a multi-center clinical trial. In particular, the randomization procedure must be updated as each patient response is received. Faries, Tamura, and Andersen found that some investigators did not always call in response data after a patient was randomized, and clinical trials personnel had to prompt investigators for missing data in order to update the randomization. Secondly, since the randomization must be dynamically updated, it is not possible to generate the randomization sequence in advance. The investigator cannot assign packaged drug sequentially and must contact the coordinating center or sponsor for the proper packages (if they are prepackaged) for each individual patient. As discussed in Section 9.6.1, since we do not know exactly the number of patients to be assigned to each treatment, there will need to be some oversupply in packaging. Faries, Tamura, and Andersen found that the system worked reasonably well in adaptive clinical trials they ran, but it required additional resources. They had two research associates on-call at all times since some investigators randomized patients on weekends and after hours and called in to get the randomization assignment.

Stratification is straightforward with response-adaptive randomization, as it is with restricted randomization procedures. One simply produces a separate randomization sequence within each of the strata. In particular, for urn models, one can run a separate urn within each stratum.

Faries, Tamura, and Andersen (1995, p. 5) concluded that

> ... We feel that the only way to gain experience [with response-adaptive randomization] is to conduct such trials and learn from our successes and failures. We encourage our clinical colleagues in the biopharmaceutical industry to do the same.

12.4   SELECTION OF A PROCEDURE

Choosing to implement a response-adaptive randomization procedure will require additional time and effort from the statistician both to select an appropriate randomization procedure and to fuel understanding by scientific colleagues in the clinical trial. In many clinical trials where there is a rush to determine an appropriate protocol, the effort required cannot be reasonably accomplished. Selection of a procedure requires simulation of the procedure under various possible clinical trial conditions. There are several aspects that the statistician should investigate:

1. Under a realistic model of the patient responses, will the response-adaptive randomization procedure work as intended? Will more patients, on average, be assigned to the superior treatment? Is the variability of the procedure within reasonable limits?

2. What is the required sample size to maintain a reasonable level of power for the study? If this sample size is larger than that required for equal allocation, are there really any savings in terms of expected numbers of treatment failures or expected numbers of patients assigned to the inferior treatment?

3. What if there is a drift in patient characteristics over time? Will this affect the adaptation adversely or introduce a covariate imbalance?

When simulating sample size and power, we have found that the easiest way is to compute the sample size n^* required for a standard clinical trial with equal allocation under an alternative reflective of the clinically relevant treatment effect, as discussed in Section 2.6. Then the response-adaptive randomization procedure is simulated k times with n^* patients, and the proportion of the k times the test statistic rejects the null hypothesis is then the simulated power of the procedure. If the procedure is less powerful than equal allocation, one then increases n^* and reruns the simulation until the power is similar. One then also simulates the expected number of treatment failures or the expected number assigned to the inferior treatment and compares this value to that obtained with equal allocation.

By using sophisticated data structures, such as priority queues, one could also incorporate delayed response into the simulation, by assuming arrivals are staggered, perhaps according to a uniform distribution, as discussed in Section 2.5, and response is delayed according to some time-to-event distribution. Patient entries and responses are then followed through a queuing system; this can be programmed using a priority queue (see, for example, Rosenberger and Seshaiyer (1997); Rosenberger and Hu (1999)).

For binary response trials with fairly immediate response, the sequential maximum likelihood procedure targeting R^*, discussed in Section 10.4.1, appears to be the most powerful procedure with the maximum savings of patients, when p_A and p_B are less than 0.5. The variability in the doubly-biased coin design has been found to be smaller than that of the sequential maximum likelihood procedure, but its impact on power has not yet been investigated. If treatments are suspected to be more successful than 0.5, then the drop-the-loser rule in Section 10.5.3 has proven to be the most powerful. The randomized play-the-winner rule is particularly variable when $p_A + p_B > 3/2$ and resultant losses in power make it unattractive. For survival outcomes, the treatment effect mapping approach has proven quite successful in terms of power and expected number of treatment failures (see Section 10.6).

12.5   BENEFITS OF RESPONSE-ADAPTIVE RANDOMIZATION

The potential benefits of adaptive allocation for clinical trials was recognized quite early. In 1969, Cornfield, Halperin, and Greenhouse (p. 760) wrote:

> Application of these results might ease the ethical problem involved in trials on human subjects. The usual ethical justification for not administering an agent of possible efficacy to all patients is the absence of definite information about its effectiveness. However satisfactory this justification may be before the trial starts it rapidly loses cogency as

evidence for or against the agent accumulates during the course of the trial. But any solution ... which permits adaptive behavior ... at least reduces this ethical problem.

Weinstein's (1974) special article in *New England Journal of Medicine* strongly advocated adaptive allocation as an alternative to traditional treatment assignment rubrics (pp. 1279, 1284):

> ... Any decision rule for allocating patients to clinical procedures in any way other than according to the best interest of the subject at hand does entail a sacrifice on the part of the subject .... Adaptive methods should be used as a matter of course. It never pays to commit oneself to a protocol under which information available before the study or obtained during its course is ignored in the treatment of a patient.

Byar, Simon, Friedewald, *et al.* (1976) responded to Weinstein's article by pointing out many of the subtle problems with adaptive designs. The comments are extremely cogent, especially in light of the limited existing literature on the subject at the time. They point out the potential for biases with time-heterogeneity, the potential loss of power due to unequal sample sizes, and the difficulty of applying the methodology to long-term trials of chronic diseases.

Other authors have argued heatedly against any form of response-adaptive *randomization*. Royall (1991, p. 58) writes:

> ... The ethical problems are clear: after finding enough evidence favoring [treatment] A to require reducing the probability of [treatment] B, the physician ... must see that the next patient gets A, not just with high probability, but with certainty.

This point was argued extensively in discussion to Royall's paper; see particularly the response of Byar. Simon (1991) writes:

> [I do] not find it attractive to approach a patient saying that I do not know which treatment is better, but treatment A is doing better therefore I will give you a greater than 50 percent chance of getting it.

While response-adaptive randomization does not eliminate the ethical problem of randomizing patients to the inferior treatment, it mitigates it by making the probability of assignment to the inferior treatment smaller. We find this to be an attractive alternative to the usual 50 : 50 randomization procedures for certain clinical trials. We believe patients and physicians will find it attractive too. Many clinical trials have used balanced allocation to multiple treatment arms as a successful recruitment tool; for example, a trial of three experimental therapies versus a placebo or a trials with a combination therapy arm, two single-therapy arms, and a placebo. In such cases one can advertise that patients have a 75 percent chance of being assigned to an experimental arm. Response-adaptive randomization can be used similarly as a recruitment tool. In truth, patients do not enter clinical trials in order to be on a placebo (although many patients would prefer to be assigned to the placebo in cases where there may be some risk of adverse events).

While Royall's point advocates deterministic assignments based on an adaptive procedure, such studies are prone to the biases of nonrandomized studies. We prefer to maintain the benefits of randomization while increasing the number of patients assigned to the superior treatment, if it exists. Tamura, Faries, Andersen, *et al.* (1994, p. 775) give the following reason for the controversy around response-adaptive randomization:

> We believe that because [response-adaptive randomization] represents a middle ground between the community benefit and the individual patient benefit, it is subject to attack from either side.

Following an adaptive clinical trial on fluoxetine for depression, (see Section 12.6.2), the investigators reported (Tamura, Faries, Andersen, *et al.* (1994, p. 775)):

> We were encouraged by the cooperation and willingness of our clinical research colleagues and our investigators to design, implement, and report on such a trial.... This has encouraged us to continue research efforts into both the implementation and analysis of adaptive trials.

12.6   SOME EXAMPLES

12.6.1   The Extracorporeal Membrane Oxygenation trial

The randomized play-the-winner rule was used in a clinical trial of extracorporeal membrane oxygenation (ECMO; Bartlett, Roloff, Cornell, *et al.*, 1985), a surgical procedure for newborns with respiratory failure. The technique had been used when infants were moribund and unresponsive to conventional treatment (ventilation and pharmacologic therapy). Early trials on safely and efficacy had indicated that the ECMO technique was safe and had an overall success rate of 56 percent, compared to a success rate of about 20 percent for conventional therapy. Bartlett, Roloff, Cornell, *et al.* (1985, p. 480) state that the $RPW(1,1)$ rule was chosen for the following reasons:

> (1) the outcome of each case [was] known soon after randomization, making it possible to use; (2) [it was] anticipated that most ECMO patients would survive and most control patients would die, so significance could be reached with a modest number of patients; [and] (3) it was a reasonable approach to the scientific/ethical dilemma.

In the randomization scheme, the first patient was assigned to ECMO and survived, changing the urn composition to 2 ECMO balls and 1 control ball. The second patient was assigned to conventional therapy and died, leading to 3 ECMO balls and 1 control ball. Each subsequent randomization was to ECMO, and each of the patients survived.

The trial was stopped after 12 total patients, using a stopping rule given described by Cornell, Landenberger, and Bartlett (1986).

Serious questions arose about the validity of such a trial. The foremost question raised is whether two treatments can adequately be compared when only one patient was assigned to one of the treatments. The validity of clinical trials with a sample size of 12 has also been questioned. In any event, the clinical trials were not convincing and led to at least two other clinical trials of the same therapy (O'Rourke, Crone, Vacanti, *et al.* (1989); see also Ware (1989); UK Collaborative ECMO Trial Group (1996)).

What went wrong? We know from Chapter 10 that the $RPW(1,1)$ rule is highly variable, particularly when $p_A + p_B > 3/2$, when the variance depends on the initial composition of the urn. In retrospect, starting with more than one ball of each type should have resulted in more patients on the control arm, and a minimum sample size should have been set in advance. To this day, some investigators use the ECMO example as a reason not to perform response-adaptive randomization at all. This is unfortunate because we think this is exactly the type of trial for which response-adaptive randomization would be particularly advantageous.

12.6.2   The fluoxetine trial

The $RPW(1,1)$ rule was employed in a clinical trial of fluoxetine versus placebo for depressive disorder. The trial was stratified by normal and shortened rapid eye movement latency (REML), so two urns were used in the randomization. In order to avoid and ECMO-like situation with too few controls, the first six patients in each stratum were assigned using a permuted block design. The primary outcome, a reduction of 50 percent or greater on the Hamilton Depression Scale between baseline and final active visit after a minimum of three weeks of therapy could only be ascertained after approximately eight weeks. Determining that this was too long a period in which to run an adaptive trial, investigators used a surrogate outcome to update the urn. They defined a surrogate responder as a patient exhibiting a reduction greater than 50 percent on the Hamilton Depression Scale in two consecutive vistis after at least three weeks of therapy. The trial was stopped after 61 patients had responded in accordance with the surrogate criterion; the trial randomized a total of 89 patients: 21 fluoxetine patients and 20 placebo patients in the shortened REML stratum; 21 fluoxetine and 21 placebo patients in the normal REML stratum. Six patients did not have a final outcome status. A significant treatment effect was found in the shortened REML category, but not the normal REML stratum. The primary outcome was analyzed using a Monte-Carlo randomization-based analysis. Although there was a significant treatment effect in the shortened REML stratum, the randomization did not favor the treatment arm. The investigators found that the randomization sequence tended to assign patients to placebo when the probability of allocation to fluoxetine was higher. They found that the probability of their particular sequence, given the allocation probabilities, was about 22 percent. (See Tamura, Faries, Andersen, *et al.*, 1994.)

12.7   CONCLUSIONS

Response-adaptive randomization procedures require more work to implement, in that the randomization procedure must be programmed and the program must update the allocation probabilities after each patient response. They also require much work on the part of the statistician in the design phase of the trial. We recommend that extensive simulations be run to ascertain the operating characteristics of the procedure, to determine sample size requirements, and to assess the potential benefits of using response-adaptive randomization. The fluoxetine trial is an example of a well-conducted and thoughtfully designed clinical trial. However, the added benefit to patients was minimal, because the allocation was close to equal even in the stratum where there was a treatment effect.

Rosenberger (1999) discusses conditions under which the use of response-adaptive randomization is reasonable. We note some of them here:

1. The therapies have been evaluated previously for toxicity. This is important to ensure that the response-adaptive randomization does not place more patients on a highly toxic treatment.
2. Delay in response is moderate, allowing the adaptation to take place.
3. Duration of the trial is limited and recruitment can take place during most or all of the trial.
4. The trial is carefully planned with extensive simulations run under different models.
5. The experimental therapy is expected to have significant benefits to the public health.
6. Modest gains in terms of treatment successes are desirable from an ethical standpoint.

Few areas of statistics have contributed to more controversy than response-adaptive randomization (see Problem 12.1). However, the extra effort required to design and implement clinical trials using response-adaptive randomization could result in significant benefits to patients and clinical medicine in general.

12.8   PROBLEMS

12.1 a. Familiarize yourself with the two famous ECMO trials by looking at the original papers (Bartlett, Roloff, Cornell, *et al.*, 1985; O'Rourke, Crone, Vacanti, *et al.*, 1989).
b. Now read about the controversy that ensued in the following papers and attendant discussions (Ware, 1989; Royall, 1991).
c. Now familiarize yourself with the 1996 UK Collaborative ECMO Trial and read the accompanying editorial (UK Collaborative ECMO Trial Group, 1996).
d. Write a 15 minute position paper to be presented in a class debate on the three ECMO trials. Focus on the following issues:
(i) Were the three trials necessary? If not, what were the appropriate alternatives?

(ii) Should an adaptive design have been used for this type of trial? If so, which one and why?

(iii) Should adaptive designs ever be used? Under what conditions?

(iv) Is randomization necessary? Are clinical trials ethical? Focus in particular on the interchange between Royall and Byar in the Royall (1991) paper.

(v) Was the 1995 UK Collaborative Trial ethical?

12.2 Find a clinical trial in a major medical journal (*e.g.*, *New England Journal of Medicine, Journal of the American Medical Association, Lancet*) for which response-adaptive randomization would be appropriate. Write a short paper explaining why this would be an appropriate trial, and describing procedures and statistical considerations in redesigning the trial using response-adaptive randomization.

12.3 For the randomization procedure in Problem 12.2, find, by simulation, the sample size necessary to attain 90 percent power for a specific alternative of interest.

12.4 For the scenario described in Section 12.2, wherein the probability of assignment to treatment A shifts from 0.5 to 0.8 in the first and second stages of recruitment, and the probability of recruiting a male shifts from 0.5 to 0.7, for equal numbers recruited in both stages show that the probability of a male is 0.623 in treatment group A and 0.557 in treatment group B.

12.9 REFERENCES

BARTLETT, R. H., ROLOFF, D. W., CORNELL, R. G., ANDREWS, A. F., DILLON, P. W., ZWISCHENBERGER, J. B. (1985). Extracorporeal circulation in neonatal respiratory failure: a prospective randomized study. *Pediatrics* **76** 479–487.

BYAR, D. P., SIMON, R. M., FRIEDEWALD, W. T., SCHLESSELMAN, J. J., DeMETS, D. L., ELLENBERG, J. H., GAIL, M. H., AND WARE, J. H. (1976). Randomized clinical trials: perspectives on some recent ideas. *New England Journal of Medicine* **295** 74–80.

CORNELL, R. G., LANDENBERGER, B. D., AND BARTLETT, R. H. (1986). Randomized play the winner clinical trials. *Communications in Statistics – Theory and Methods* **15** 159–178.

CORNFIELD, J., HALPERIN, M., GREENHOUSE, S. W. (1969). An adaptive procedure for sequential clinical trials. *Journal of the American Statistical Association* **64** 759–770.

FARIES, D. E., TAMURA, R. N., AND ANDERSEN, J. S. (1995). Adaptive designs in clinical trials: Lilly experience. *Biopharmaceutical Report* **3:1** 1–11, with discussion.

O'ROURKE, P. P., CRONE, R. K., VACANTI, J. P., WARE, J. H., LILLE-HEI, C. W., PARAD, R. B., EPSTEIN, M. F. (1989). Extracorporeal membrane oxygenation and conventional medical therapy in neonates with persistent pulmonary hypertension of the newborn: a prospective random-ized study. *Pediatrics* **84** 957–963.

ROSENBERGER, W. F. (1996). New directions in adaptive designs. *Statistical Science* **11** 137–149.

ROSENBERGER, W. F. (1999). Randomized play-the-winner clinical trials: review and recommendations. *Controlled Clinical Trials* **20** 328–342.

ROSENBERGER, W. F. AND HU, F. (1999). Bootstrap methods for adaptive designs. *Statistics in Medicine* **18** 1757–1767.

ROSENBERGER, W. F. AND LACHIN, J. M. (1993). The use of response-adaptive designs in clinical trials. *Controlled Clinical Trials* **14** 471–484.

ROSENBERGER, W. F. AND SESHAIYER, P. (1997). Adaptive survival trials. *Journal of Biopharmaceutical Statistics* **7** 617–624.

ROUT, C. C., ROCKE, D. A., LEVIN, J., GOUWS, E., AND REDDY, D. (1993). A reevaluation of the role of crystalloid preload in the prevention of hypotension associated with spinal anesthesia for elective Cesarean section. *Anesthesiology* **79** 262–269.

ROYALL, R. M. (1991). Ethics and statistics in randomized clinical trials. *Statistical Science* **6** 52–62, with discussion.

SIMON, R. (1991). A decade of progress in statistical methodology for clinical trials. *Statistics in Medicine* **10** 1789–1817.

TAMURA, R. N., FARIES, D. E., ANDERSEN, J. S., AND HEILIGENSTEIN, J. H. (1994). A case study of an adaptive clinical trial in the treatment of out-patients with depressive disorder. *Journal of the American Statistical Association* **89** 768–776.

UK COLLABORATIVE ECMO TRIAL GROUP. (1996). Collaborative randomized trial of neonatal extracorporeal membrane oxygenation. *Lancet* **348** 75–82.

WARE, J. H. (1989). Investigating therapies of potentially great benefit: ECMO. *Statistical Science* **4** 298–340, with discussion.

WEINSTEIN, M. C. (1974). Allocation of subjects in medical experiments. *New England Journal of Medicine* **291** 1278–1285.

13

Some Useful Results in Large Sample Theory

13.1 SOME USEFUL CENTRAL LIMIT THEOREMS

In this chapter we provide some useful large sample theory from the probability literature that will be useful Chapters 14 and 15.* In later sections we will deal with martingale theory which is necessary to deal with the dependence structure of most of the ranodmization procedures. In this section, we begin with some useful central limit theorems that do not require martingale theory.

The linear rank statistic under complete randomization is a sum of independent, but not identically distributed, random variables. The usual central limit theorem dealing with this case is the Lindeberg-Feller Central Limit Theorem (*e.g.*, Laha and Rohatgi (1979, p. 282)).

Lindeberg-Feller Theorem. Let $\{X_n\}$ be a sequence of independent random variables with $E(X_n) = \alpha_n$ and $\text{Var}(X_n) = \sigma_n^2 < \infty, n = 1, 2, \ldots$. Let $S_n = \sum_{j=1}^{n} X_j$ and let F_n be the distribution function of X_n. If the following condition holds for every

---

*The prerequisite for this chapter is a graduate-level probability text, such as Chung (1974). With the exception of the well-known Lindeberg-Feller Central Limit Theorem, there is no convention on the names of the theorems in this chapter. We simply label them according to the literature from which they were extracted, for convenience in later chapters. Such a labeling scheme may ignore the historical roots and originators of the theory on which these theorems are based.

$\epsilon > 0$:

$$\lim_{n \to \infty} \frac{1}{\sum_{j=1}^{n} \sigma_j^2} \sum_{k=1}^{n} \int_{|x - \alpha_k| \geq \epsilon \sqrt{\sum_{j=1}^{n} \sigma_j^2}} (x - \alpha_k)^2 dF_k(x) = 0, \qquad (13.1)$$

then

$$\frac{S_n - E(S_n)}{\sqrt{\sum_{j=1}^{n} \sigma_j^2}} \to N(0, 1), \qquad (13.2)$$

in law, as $n \to \infty$.

We can use the Lindeberg-Feller theorem to derive the asymptotic distribution of the *unconditional* linear rank test under complete randomization. For the conditional test, we need a conditional central limit theorem (Holst, 1979). The following is a adaptation of Holst's Theorem 3 (p. 555) that is useful in Chapter 14.

Holst's Theorem. Let $(a_{1n}, ..., a_{nn})$ be real constants satisfying the following:

$$\sum_{j=1}^{n} a_{jn} = 0, \qquad (13.3)$$

$$\frac{\sum_{j=1}^{n} a_{jn}}{n} \to \gamma, \qquad (13.4)$$

and

$$\frac{\max_{1 \leq j \leq n} a_{jn}^2}{n} \to 0. \qquad (13.5)$$

Let $\{X_n\}$ be independent and identically distributed random variables with $E(X) = \theta$ and $\text{Var}(X) = \sigma^2(\theta)$ satisfying the following:

$$\sum_{j=1}^{n} X_j \text{ is sufficient for } \theta \qquad (13.6)$$

and for every $\epsilon > 0$ there exists $K_\epsilon < 1$ such that, for $\epsilon \leq |t| < 1$,

$$\left| E\left(e^{it(X_j - \theta)} \right) \right| \leq K_\epsilon < 1. \qquad (13.7)$$

If (13.3)–(13.7) hold, and $x_n/n \to \theta$ as $n \to \infty$, then

$$n^{-1/2} \frac{\sum_{j=1}^{n} a_{jn} X_j}{\gamma^{1/2} \sigma(x_n/n)},$$

conditional on $\sum_{j=1}^{n} X_j = x_n$, converges in law to a standard normal variate as $n \to \infty$.

The third central limit theorem of interest is an adaptation of Theorem 4A from Hájek (1969, p. 14). It is useful in proving asymptotic normality of the linear rank test when the random allocation rule is employed.

Hájek's Theorem. Let $\{R_{1n}, ..., R_{nn}\}$ have a uniform distribution over the n, permutations of $(1, ..., n)$. Consider a statistic of the form

$$S_n = \sum_{j=1}^{n} c_{jn} \left(u_n + v_n \psi \left(\frac{R_{jn}}{n+1} \right) \right), \tag{13.8}$$

where $v_n \neq 0$, ψ is nondecreasing on $(0,1)$, and $\int_0^1 (\psi(t) - \bar{\psi})^2 dt \in (0, \infty)$ ($\bar{\psi} = \int_0^1 \psi(t)dt$). Then if

$$\lim_{n\to\infty} \frac{\max\limits_{1\le j\le n} (c_{jn} - \bar{c}_n)^2}{\sum_{j=1}^{n}(c_{jn} - \bar{c}_n)^2} = 0,$$

where $\bar{c}_n = n^{-1} \sum_{i=1}^{n} c_{in}$, we have

$$\frac{S_n - E(S_n)}{\sqrt{\text{Var}(S_n)}} \to N(0,1), \tag{13.9}$$

in law, as $n \to \infty$.

13.2 MARTINGALES AND SUMS OF DEPENDENT RANDOM VARIABLES

In a general sense, let $\{\mathcal{F}_n\}$ be an increasing sequence of sigma-algebras and let $Z_n \in \mathcal{F}_n$ be a sequence of random variables such that $E|Z_n| < \infty$. Then if

$$E(Z_n|\mathcal{F}_m) = Z_m, \text{ almost surely, for all } m < n, \tag{13.10}$$

then Z_n is a martingale with respect to $\mathcal{F}_n$ (Hall and Heyde (1980, p. 1)). In particular, since the sigma-algebras are nested, we can rewrite (13.10) as

$$E(Z_n|\mathcal{F}_{n-1}) = Z_{n-1} \text{ almost surely.} \tag{13.11}$$

In most of our applications, we are concerned with sums of *dependent* random variables, so that $S_n = \sum_{j=1}^{n} X_j$ and $\mathcal{F}_n = \sigma\{X_1, ..., X_n\}$. Assume that $E|X_n| < \infty$ for all n. Note that $\mathcal{F}_n$ is increasing. Any sum of dependent random variables can be transformed into a martingale by letting $Z_n = \sum_{j=1}^{n}(X_j - E(X_j|\mathcal{F}_{j-1}))$. This is because

$$
\begin{aligned}
E(Z_n|\mathcal{F}_{n-1}) &= E\left(\sum_{j=1}^{n}(X_j - E(X_j|\mathcal{F}_{j-1}))\middle|\mathcal{F}_{n-1}\right) \\
&= E(X_n - E(X_n|\mathcal{F}_{n-1})|\mathcal{F}_{n-1}) \\
&\quad + E\left(\sum_{j=1}^{n-1}(X_j - E(X_j|\mathcal{F}_{j-1}))\middle|\mathcal{F}_{n-1}\right) \\
&= E(X_n|\mathcal{F}_{n-1}) - E(X_n|\mathcal{F}_{n-1}) + \sum_{j=1}^{n-1}(X_j - E(X_j|\mathcal{F}_{i-1})) \\
&= Z_{n-1}.
\end{aligned}
$$

We thus gain two important insights into sums of dependent random variables. First, to verify if $\sum_{j=1}^{n} X_j$ is a martingale, it suffices to show only that $E(X_n|\mathcal{F}_{n-1}) = 0$. Second, any sum of dependent random variables can be turned into a martingale by subtracting its *compensator*, $E(X_j|\mathcal{F}_{j-1})$ termwise. It is easy to see, for instance, that sums of independent zero-mean random variables form a martingale, as $E(X_n|\mathcal{F}_{n-1}) = E(X_n) = 0$.

We are interested in martingales because similar limit theorems as for sums of independent random variables are available in the literature, such as the weak law of large number and the central limit theorem. For the martingale weak law, let $S_n = \sum_{j=1}^{n} X_j$ be a zero-mean martingale (*i.e.*, already compensated). Then if

$$
\frac{\sum_{j=1}^{n} E(X_j^2)}{n^2} \to 0, \tag{13.12}
$$

we have

$$
\frac{S_n}{n} \to 0 \tag{13.13}
$$

in probability. To see this, simply use Chebyshev's inequality:

$$
\Pr(|S_n| > \epsilon) \le \frac{E(S_n^2)}{\epsilon^2}.
$$

For the right-hand side,

$$
E(S_n^2) = \sum_{j=1}^{n} E(X_j^2) + 2\sum_{i>j}(X_i X_j).
$$

For $i > j$,

$$
\begin{aligned}
E(X_i X_j) &= EE(X_i X_j|\mathcal{F}_{i-1}) \\
&= E(X_j E(X_i|\mathcal{F}_{i-1})) = 0.
\end{aligned}
$$

So

$$\Pr(|S_n| \geq n\epsilon) \leq \frac{\sum_{j=1}^{n} E(X_j^2)}{\epsilon^2 n^2} \to 0$$

by (13.12), and hence (13.13) holds (Hall and Heyde, 1980).

Billingsley (1961, p. 52) provides a very simple central limit theorem for sums of dependent random variables.

Billingsley's Theorem. Let $S_n = \sum_{j=1}^{n} X_j$ be a zero mean martingale with respect to $\mathcal{F}_n$. Under the following two conditions:

$$\lim_{n \to \infty} n^{-1} \sum_{j=1}^{n} E(X_j^2 | \mathcal{F}_{j-1}) = \beta^2 \tag{13.14}$$

and

$$\lim_{n \to \infty} n^{-1-\delta/2} \sum_{j=1}^{n} E(|X_j|^{2+\delta} | \mathcal{F}_{j-1}) = 0 \tag{13.15}$$

almost surely, for positive constants δ and β^2, then

$$n^{-1/2} \sum_{j=1}^{n} X_j \to N(0, \beta^2), \tag{13.16}$$

in law, as $n \to \infty$.

Another useful result in this context is the Cramér-Wold Device (*e.g.*, Billingsley, 1968), which allows us to extend central limit results from scalars to vectors.

The Cramér-Wold Device. Define vectors of random variables $\boldsymbol{X}_n = (X_{n1}, ..., X_{ns})$ and $\boldsymbol{X} = (X_1, ..., X_s)$. If

$$\sum_{i=1}^{s} \alpha_i X_{ni} \to \sum_{i=1}^{s} \alpha_i X_i$$

in law, as $n \to \infty$, for any arbitrary sequence of constants $\alpha_1, ..., \alpha_s$, then $\boldsymbol{X}_n \to \boldsymbol{X}$ in law.

13.3 MARTINGALES AND TRIANGULAR ARRAYS

Billingsley's Theorem is not broad enough for some of our purposes. We require, in particular for proving the asymptotic normality of linear rank statistics, a martingale central limit theorem for triangular arrays. For each $j = 1, ..., n$, let S_{nk} be a martingale with respect to nested sigma-algebras $\mathcal{F}_{nk}$. (By nested we mean $\mathcal{F}_{nk} \subseteq \mathcal{F}_{n+1,k}, k = 1, ..., n, n \geq 1$.) Let $X_{nk} = S_{nk} - S_{n,k-1}, S_{n0} = 0$, denote the martingale differences. Then $\{S_{nk}, \mathcal{F}_{nk}\}$, for $k = 1, ..., n, n \geq 1$ is a double sequence of triangular arrays, called a *martingale array* (Hall and Heyde (1980,

p. 52)). For example, in (14.12), W_{kn} forms a martingale array with $\mathcal{F}_{nk} = \sigma\{T_1, ..., T_k\}$.

We have the following central limit theorem for martingale arrays $\{S_{nk}, \mathcal{F}_{nk}\}$ (Hall and Heyde (1980, p. 58)).

Hall and Heyde's Theorem. Let η^2 be a positive constant. If the following three conditions hold:

$$\max_{1 \leq k \leq n} |X_{nk}| \to 0 \tag{13.17}$$

in probability, as $n \to \infty$;

$$\sum_{k=1}^{n} X_{nk}^2 \to \eta^2, \tag{13.18}$$

in probability, as $n \to \infty$;

$$E(\max_{1 \leq k \leq n} X_{nk}^2) \tag{13.19}$$

is bounded in n, then

$$S_{nn} \to N(0, \eta^2), \tag{13.20}$$

in law, as $n \to \infty$. The theorem also holds if we replace (13.17) and (13.19) with

$$\sum_{k=1}^{n} E(X_{nk}^2 I(|X_{nk}| > \epsilon)|\mathcal{F}_{n,k-1}) \to 0 \tag{13.21}$$

in probability, as $n \to \infty$, for all $\epsilon > 0$, and we replace (13.18) with

$$\sum_{k=1}^{n} E(X_{nk}^2|\mathcal{F}_{n,k-1}) \to \eta^2, \tag{13.22}$$

in probability, as $n \to \infty$.

13.4 ASYMPTOTIC NORMALITY OF MAXIMUM LIKELIHOOD ESTIMATORS

We can use the weak law of large numbers and central limit theorem for martingales to show consistency and asymptotic normality of the maximum likelihood estimators for dependent data (see Hall and Heyde (1980, Section 6.2) for the scalar case). In particular, this is useful in Chapter 15, where we wish prove the asymptotic normality of maximum likelihood estimators from a response-adaptive randomization procedure.

13.4.1   The likelihood

Following Rosenberger, Durham, and Flournoy (1997), let $\boldsymbol{\theta}$ be a vector-valued parameter of s dimensions. Define $\mathcal{L}_n(\boldsymbol{\theta}) = \Pr_n(x_1, ..., x_n; \boldsymbol{\theta})$ to be the likelihood of the data $(x_1, ..., x_n)$, where $\mathcal{L}_0 = 1$. Assume that $\Pr_n(x_1, ..., x_n; \boldsymbol{\theta})$ can be partially differentiated twice (with respect to $\boldsymbol{\theta}$) and that integration and differentiation can be interchanged. Then the first derivative of the loglikelihood can be written as

$$\frac{\partial \log \mathcal{L}_n(\boldsymbol{\theta})}{\partial \theta_j} = \sum_{i=1}^{n} \frac{\partial}{\partial \theta_j} (\log \mathcal{L}_i(\boldsymbol{\theta}) - \log \mathcal{L}_{i-1}(\boldsymbol{\theta}))$$

$$= \sum_{i=1}^{n} \frac{\partial}{\partial \theta_j} L_i(\boldsymbol{\theta}), \qquad (13.23)$$

$j = 1, ..., s$, where $L_i(\boldsymbol{\theta}) = \log(\mathcal{L}_i(\boldsymbol{\theta})/\mathcal{L}_{i-1}(\boldsymbol{\theta}))$. Let $\mathcal{F}_n$ be the sigma algebra generated by the stochastic process through stage n, with $\mathcal{F}_0$ the trivial sigma algebra.

We first show that the first derivative of the loglikelihood is a martingale with respect to $\mathcal{F}_n$. Since it can be expressed as a sum of dependent random variables via (13.23), we need show only that

$$E\left(\frac{\partial}{\partial \theta_j} L_n(\boldsymbol{\theta}) \middle| \mathcal{F}_{n-1}\right) = 0.$$

This follows since

$$E\left(\frac{\partial}{\partial \theta_j} L_n(\boldsymbol{\theta}) \middle| \mathcal{F}_{n-1}\right) = E\left(\log(\mathcal{L}_n(\boldsymbol{\theta})/\mathcal{L}_{n-1}(\boldsymbol{\theta})) \middle| \mathcal{F}_{n-1}\right)$$

$$= \int \frac{\partial}{\partial \theta_j} \log\left(\frac{\Pr_n(x_1, ..., x_n; \boldsymbol{\theta})}{\Pr_{n-1}(x_1, ..., x_{n-1}; \boldsymbol{\theta})}\right) \frac{\Pr_n(x_1, ..., x_n; \boldsymbol{\theta})}{\Pr_{n-1}(x_1, ..., x_{n-1}; \boldsymbol{\theta})} dx_n$$

$$= \frac{\int \frac{\partial}{\partial \theta_j}(\Pr_n(x_1, ..., x_n; \boldsymbol{\theta})) \Pr_{n-1}(x_1, ..., x_{n-1}; \boldsymbol{\theta}) dx_n}{(\Pr_{n-1}(x_1, ..., x_{n-1}; \boldsymbol{\theta}))^2}$$

$$- \frac{\int \frac{\partial}{\partial \theta_j}(\Pr_{n-1}(x_1, ..., x_{n-1}; \boldsymbol{\theta})) \Pr_n(x_1, ..., x_n; \boldsymbol{\theta}) dx_n}{(\Pr_{n-1}(x_1, ..., x_{n-1}; \boldsymbol{\theta}))^2}$$

$$= \frac{\frac{\partial}{\partial \theta_j}\left(\int \Pr_n(x_1, ..., x_n; \boldsymbol{\theta}) dx_n\right)}{\Pr_{n-1}(x_1, ..., x_{n-1}; \boldsymbol{\theta})}$$

$$- \frac{\frac{\partial}{\partial \theta_j}(\Pr_{n-1}(x_1, ..., x_{n-1}; \boldsymbol{\theta})) \int \Pr_n(x_1, ..., x_n; \boldsymbol{\theta}) dx_n}{(\Pr_{n-1}(x_1, ..., x_{n-1}; \boldsymbol{\theta}))^2}$$

$$= \frac{\frac{\partial}{\partial \theta_j}(\Pr_{n-1}(x_1, ..., x_{n-1}; \boldsymbol{\theta}))}{\Pr_{n-1}(x_1, ..., x_{n-1}; \boldsymbol{\theta})} - \frac{\frac{\partial}{\partial \theta_j}(\Pr_{n-1}(x_1, ..., x_{n-1}; \boldsymbol{\theta}))}{\Pr_{n-1}(x_1, ..., x_{n-1}; \boldsymbol{\theta})}$$

$$= 0.$$

Since the first derivative is a martingale, and is just a sum of dependent random variables, we can use the weak law and Billingsley's Theorem from Section 13.2 to prove consistency and asymptotic normality of the maximum likelihood estimator $\hat{\boldsymbol{\theta}}$. The proof follows basically the same format as for the independent and identically distributed case, and the reader is referred to Lehmann (1983, Section 6.4) for details. Essentially, the likelihood is expanded in a Taylor series, with the first derivative of the loglikelihood, suitably normalized, converging in law to a normal distribution and the second derivatives converging to Fisher's information.

13.4.2   Basic conditions for consistency and asymptotic normality

Assume the following standard regularity conditions hold:

1. There exists an open subset ω of the parameter space Ω containing the true parameter $\boldsymbol{\theta}^0$.
2. The first partial derivatives of the loglikelihood have finite moments of order $2 + \delta$ for some $\delta > 0$.
3. The likelihood $\mathcal{L}_n(\boldsymbol{\theta})$ admits all third partial derivatives, and the absolute values of the third partials (with respect to θ_j, θ_k, and θ_l) are bounded by functions $M_{jkl}(x_1, ..., x_n)$ for all $\boldsymbol{\theta} \in \omega$, where $E_{\boldsymbol{\theta}^0}(M_{jkl}(X_1, ..., X_n)) < \infty$.

These regularity conditions will be trivially satisfied for the applications in this book.

Under the following three conditions, for $j = 1, ..., s, k = 1, ..., s,$:

$$n^{-1} \sum_{i=1}^{n} \frac{\partial}{\partial \theta_j} L_i(\boldsymbol{\theta}) \to 0, \tag{13.24}$$

in probability, for all $\boldsymbol{\theta} \in \omega$, as $n \to \infty$;

$$n^{-1/2} \sum_{i=1}^{n} \frac{\partial}{\partial \theta_j} L_i(\boldsymbol{\theta}) \to N_s(0, \Gamma(\boldsymbol{\theta})), \tag{13.25}$$

in law, for all $\boldsymbol{\theta} \in \omega$, as $n \to \infty$, where $\Gamma(\boldsymbol{\theta}) = ((\gamma_{jk}))$;

$$n^{-1} \sum_{i=1}^{n} \frac{\partial^2}{\partial \theta_j \theta_k} L_i(\boldsymbol{\theta}) \to -\gamma_{jk}, \tag{13.26}$$

in probability, for all $\boldsymbol{\theta} \in \omega$, as $n \to \infty$, we have that a consistent maximum likelihood estimator $\hat{\boldsymbol{\theta}}$ exists, and the vector (for $j = 1, ..., s$) given by

$$n^{1/2}(\hat{\theta}_j - \theta_j)$$

is asymptotically multivariate normal with mean zero and variance-covariance matrix $(\Gamma(\boldsymbol{\theta}))^{-1}$, provided the inverse exists (Rosenberger, Flournoy, and Durham, 1997).

13.4.3   Alternative conditions

For most applications, it is not necessary to show (13.24)-(13.26) directly, and it is instructive to explore each of these conditions further. Let us first examine condition

(13.24), which is just a weak law for martingales. From (13.12) and (13.13), we require that

$$n^{-2} \sum_{i=1}^{n} E\left(\frac{\partial}{\partial \theta_j} L_i(\boldsymbol{\theta})\right)^2 \to 0. \tag{13.27}$$

Note that this is trivially satisfied if the first derivatives are bounded in i.

For condition (13.25), we can apply Billingsley's Theorem and the Cramér-Wold Device from Section 13.2. In fact, we now show that, under the following two conditions, (13.25) holds:

$$n^{-1} \sum_{i=1}^{n} E\left(\frac{\partial}{\partial \theta_j} L_i(\boldsymbol{\theta}) \frac{\partial}{\partial \theta_k} L_i(\boldsymbol{\theta}) \Big| \mathcal{F}_{i-1}\right) \to \gamma_{jk}, \tag{13.28}$$

almost surely, as $n \to \infty$; and

$$n^{-1-\delta/2} \sum_{i=1}^{n} E\left(\left|\frac{\partial}{\partial \theta_j} L_i(\boldsymbol{\theta})\right|^{2+\delta} \Big| \mathcal{F}_{i-1}\right) \to 0, \tag{13.29}$$

almost surely, as $n \to \infty$.

To show this, define arbitrary constants $\alpha_1, ..., \alpha_s$. Then

$$\sum_{i=1}^{n} \sum_{j=1}^{s} \alpha_j \frac{\partial}{\partial \theta_j} L_i(\boldsymbol{\theta})$$

is a martingale. Checking condition (13.14) of Billingsley's Theorem, we have

$$n^{-1} \sum_{i=1}^{n} E\left(\left[\sum_{j=1}^{s} \alpha_j \frac{\partial}{\partial \theta_j} L_i(\boldsymbol{\theta})\right]^2 \Big| \mathcal{F}_{i-1}\right)$$

$$= n^{-1} \sum_{i=1}^{n} \sum_{j=1}^{s} \sum_{k=1}^{s} \alpha_j \alpha_k E\left(\frac{\partial}{\partial \theta_j} L_i(\boldsymbol{\theta}) \frac{\partial}{\partial \theta_k} L_i(\boldsymbol{\theta}) \Big| \mathcal{F}_{i-1}\right)$$

$$\to \sum_{j=1}^{s} \sum_{k=1}^{s} \alpha_j \alpha_k \gamma_{jk}$$

almost surely, as $n \to \infty$, by (13.28). Checking condition (13.15), we see that

$$n^{-1-\delta/2} \sum_{i=1}^{n} E\left(\left|\sum_{j=1}^{s} \alpha_j \frac{\partial}{\partial \theta_j} L_i(\boldsymbol{\theta})\right|^{2+\delta} \Big| \mathcal{F}_{i-1}\right)$$

$$\le n^{-1-\delta/2} \sum_{i=1}^{n} E\left(\left[\sum_{j=1}^{s} \alpha_j^2\right]^{1+\delta/2} \left[\sum_{j=1}^{s} \left(\frac{\partial}{\partial \theta_j} L_i(\boldsymbol{\theta})\right)^2\right]^{1+\delta/2} \Big| \mathcal{F}_{i-1}\right)$$

$$= K n^{-1-\delta/2} \sum_{i=1}^{n} E\left(\left[\sum_{j=1}^{s} \left(\frac{\partial}{\partial \theta_j} L_i(\boldsymbol{\theta})\right)^2\right]^{1+\delta/2} \Big| \mathcal{F}_{i-1}\right),$$

where

$$K = \left[\sum_{j=1}^{s} \alpha_j^2 \right]^{1+\delta/2},$$

by the Cauchy-Schwartz inequality. We have the following inequality from Chung (1974, p. 48). If $p > 1$,

$$\left| n^{-1} \sum_{i=1}^{n} X_i \right|^p \leq n^{-1} \sum_{i=1}^{n} |X_i|^p.$$

Using this inequality, we see that

$$Kn^{-1-\delta/2} \sum_{i=1}^{n} E\left(\left[\sum_{j=1}^{s} \left(\frac{\partial}{\partial \theta_j} L_i(\boldsymbol{\theta}) \right)^2 \right]^{1+\delta/2} \Bigg| \mathcal{F}_{i-1} \right)$$

$$\leq Ks^{\delta/2} \sum_{j=1}^{s} n^{-1-\delta/2} \sum_{i=1}^{n} E\left(\left| \frac{\partial}{\partial \theta_j} L_i(\boldsymbol{\theta}) \right|^{2+\delta} \Bigg| \mathcal{F}_{i-1} \right)$$

$$\to 0$$

almost surely, as $n \to \infty$, by (13.29). Thus, by Billingsley's Theorem,

$$n^{-1/2} \sum_{i=1}^{n} \sum_{j=1}^{s} \alpha_j \frac{\partial}{\partial \theta_j} L_i(\boldsymbol{\theta}) \to N\left(0, \sum_{j=1}^{s} \sum_{k=1}^{s} \alpha_j \alpha_k \gamma_{jk} \right),$$

in law. Employing the Cramér-Wold Device, we see that (13.25) holds.

Now let us explore condition (13.26) further. First we define

$$\mathrm{Var}(X_i | \mathcal{F}_{i-1}) = E\left((X_i - E(X_i | \mathcal{F}_{i-1}))^2 \Big| \mathcal{F}_{i-1} \right).$$

Now consider the following two conditions:

$$n^{-2} \sum_{i=1}^{n} \mathrm{Var}\left(\frac{\partial^2}{\partial \theta_j \partial \theta_k} L_i(\boldsymbol{\theta}) \right) \to 0 \qquad (13.30)$$

as $n \to \infty$ and

$$n^{-2} \sum_{i=1}^{n} E\left(\mathrm{Var}\left(\frac{\partial^2}{\partial \theta_j \partial \theta_k} L_i(\boldsymbol{\theta}) \Big| \mathcal{F}_{i-1} \right) \right) \to 0 \qquad (13.31)$$

as $n \to \infty$. It can be shown (Problems 13.3 and 13.4) that (13.30) implies (13.31) and that (13.28), together with (13.31), implies (13.26).

13.4.4 Conclusions

From results of Section 13.4.3, we have the following theorem:

Rosenberger, Flournoy, and Durham's Theorem. Assume the three regularity conditions of Section 13.4.2 hold. Then if either the set of conditions (13.27), (13.28), (13.29), and (13.26) or the set of conditions (13.27), (13.28), (13.29), and (13.30) hold, we have that a consistent estimator $\hat{\boldsymbol{\theta}}$ exists and

$$n^{1/2}(\hat{\boldsymbol{\theta}} - \boldsymbol{\theta}) \to N_s(\mathbf{0}, \boldsymbol{\Gamma}^{-1}), \tag{13.32}$$

provided $\boldsymbol{\Gamma}^{-1}$ exists. Furthermore, if the first and second partial derivatives are bounded in i for each $j = 1, ..., s$, we require only condition (13.28) for (13.32) to hold.

The last sentence is an important addendum which will be put to good use in Chapter 15. Finally, a substitute condition for (13.28) can be gleaned from Problem 13.2.

13.5 PROBLEMS

13.1 Show the equivalence of (13.10) and (13.11).

13.2 Assume that $\mathrm{Pr}_n(x_1, ..., x_n; \boldsymbol{\theta})$ can be partially differentiated twice (with respect to $\boldsymbol{\theta}$) and that integration and differentiation can be interchanged. Show that

$$E\left(\frac{\partial}{\partial \theta_j} \mathcal{L}_n(\boldsymbol{\theta}) \frac{\partial}{\partial \theta_k} \mathcal{L}_n(\boldsymbol{\theta}) \bigg| \mathcal{F}_{n-1}\right) = -E\left(\frac{\partial^2}{\partial \theta_j \partial \theta_k} \mathcal{L}_n(\boldsymbol{\theta}) \bigg| \mathcal{F}_{n-1}\right).$$

13.3 Show that (13.30) implies (13.31).

13.4 Show that (13.28), together with (13.31), implies (13.26). (*Hint:* Find an appropriate compensator and apply the weak law for martingales.)

13.6 REFERENCES

BILLINGSLEY, P. (1961). *Statistical Inference for Markov Processes.* University of Chicago Press, Chicago.

BILLINGSLEY, P. (1968). *Convergence of Probability Measures.* Wiley, New York.

CHUNG, K. L. (1974). *A Course in Probability Theory.* Academic Press, San Diego.

HÁJEK, J. (1969). *A Course in Nonparametric Statistics.* Holden-Day, San Francisco.

HALL, P. AND HEYDE, C. C. (1980). *Martingale Limit Theory and Its Application.* Academic Press, San Diego.

HOLST, L. (1979). Two conditional limit theorems with applications. *Annals of Statistics* **7** 551–557.

LAHA, R. G. AND ROHATGI, V. K. (1979). *Probability Theory.* Wiley, New York.

LEHMANN, E. L. (1983). *The Theory of Point Estimation.* Wiley, New York.

ROSENBERGER, W. F., FLOURNOY, N., AND DURHAM, S. D. (1977). Asymptotic normality of maximum likelihood estimators from multiparameter response-driven designs. *Journal of Statistical Planning and Inference* **60** 69–76.

14

Large Sample Inference for Complete and Restricted Randomization

14.1 INTRODUCTION

In this chapter, we examine the large-sample distribution of linear rank tests under the randomization procedures discussed in Chapter 3. This distribution will necessarily be determined by two components: (i) a condition on the scores $\{a_{jn}\}$ and (ii) the particular randomization procedure used.

As in Chapter 4, it will be convenient to redefine the treatment assignments by letting $T_j = 1$ or -1, $j = 1, ..., n$ to indicate treatments A and B, respectively. Note that this differs from the development in Section 7.9, where the treatment assignments are coded as $1/0$. For the most part, this only makes a difference by a constant multiple of 2, but the theoretical developments in this chapter are simpler using the $1/-1$ coding. Then we define (slightly different from Chapter 7) the linear rank test under a randomization model as

$$S_n = \sum_{j=1}^{n}(a_{jn} - \bar{a}_n)T_j. \tag{14.1}$$

Under the usual central limit arguments, one might expect that

$$\frac{S_n}{\sqrt{\text{Var}(S_n)}} \tag{14.2}$$

would converge in law to a standard normal distribution. This is often, but not always, the case. In the case of complete randomization, $\Sigma_T = I$, and the independence of $T_1, ..., T_n$ can be used with standard central limit theory arguments to obtain the an asymptotic normal distribution. However, for restricted randomization procedures, Σ_T is not diagonal, and so S_n represents the sum of *dependent* random variables. In some cases, we will have to rely on *martingale theory* (see Chapter 13 for prerequisite material) in order to prove normality of sums of dependent random variables. We now explore the asymptotic distribution of linear rank tests for each of the randomization procedures given in Chapter 3. We also use simulation to verify the asymptotic distribution for $n = 50$. It is known that permutation tests tend to be conservative when restricted randomization is employed (see Kalish and Begg, 1987).

14.2 COMPLETE RANDOMIZATION

14.2.1 The unconditional test

For complete randomization, (14.1) represents a sum of independent, but not identically distributed random variables

$$(a_{1n} - \bar{a}_n)T_1, ..., (a_{nn} - \bar{a}_n)T_n.$$

Since the a_{jn} are assumed constants and $\mathrm{Var}(T_j) = 1$,

$$\mathrm{Var}((a_{jn} - \bar{a}_n)T_j) = (a_{jn} - \bar{a}_n)^2.$$

Then

$$\mathrm{Var}(S_n) = \sum_{j=1}^{n}(a_{jn} - \bar{a}_n)^2.$$

So one would conjecture that

$$\frac{\sum_{j=1}^{n}(a_{jn} - \bar{a}_n)T_j}{\sqrt{\sum_{j=1}^{n}(a_{jn} - \bar{a}_n)^2}} \to N(0, 1) \tag{14.3}$$

in law, as $n \to \infty$.

We can use the Lindeberg-Feller central limit theorem (Section 13.1) to prove asymptotic normality of the linear rank test under the assumption that

$$\lim_{n\to\infty} \frac{\max\limits_{1\leq j\leq n}(a_{jn} - \bar{a}_n)^2}{\sum_{j=1}^{n}(a_{jn} - \bar{a}_n)^2} = 0. \tag{14.4}$$

Asymptotic normality is then assured by the Lindeberg condition, given in (13.1), which can be written as

$$\lim_{n\to\infty} \frac{1}{\sum_{j=1}^{n}(a_{jn} - \bar{a}_n)^2} \sum_{j=1}^{n}\int_{x^2\geq\epsilon^2\sum_{j=1}^{n}(a_{jn}-\bar{a}_n)^2} x^2\,dF_j(x) \to 0, \tag{14.5}$$

for all $\epsilon > 0$, where $F_j(x)$ is the distribution function of $(a_{jn} - \bar{a}_n)T_j$. To show this, the left-hand side of (14.5) is

$$
\leq \lim_{n \to \infty} \frac{\max_{1 \leq j \leq n}(a_{jn} - \bar{a}_n)^2}{\sum_{j=1}^{n}(a_{jn} - \bar{a}_n)^2} \sum_{j=1}^{n} \int_{x^2 \geq \epsilon^2 \sum_{j=1}^{n}(a_{jn} - \bar{a}_n)^2} dF_j(x)
$$

$$
= \lim_{n \to \infty} \frac{\max_{1 \leq j \leq n}(a_{jn} - \bar{a}_n)^2}{\sum_{j=1}^{n}(a_{jn} - \bar{a}_n)^2} \sum_{j=1}^{n} I\left(\frac{(a_{jn} - \bar{a}_n)^2}{\sum_{j=1}^{n}(a_{jn} - \bar{a}_n)^2} \geq \epsilon\right),
$$

where I is the indicator function. From (14.4), for all $\epsilon > 0$, there exists an $n_0(\epsilon)$ such that for $n > n_0$,

$$
I\left(\frac{(a_{jn} - \bar{a}_n)^2}{\sum_{j=1}^{n}(a_{jn} - \bar{a}_n)^2} \geq \epsilon\right) = 0
$$

for all j. This implies that

$$
\sum_{j=1}^{n} I\left(\frac{(a_{jn} - \bar{a}_n)^2}{\sum_{j=1}^{n}(a_{jn} - \bar{a}_n)^2} \geq \epsilon\right) \to 0
$$

as $n \to \infty$. Hence (14.5) holds. By (13.2), we thus obtain that, under complete randomization, (14.3) holds, provided (14.4) holds.

The simple ranks and other considerably more complicated ranking schemes, such as the van der Waarden scores (defined in Problem 7.4) satisfy (14.4) (Problem 14.2). From Smythe and Wei (1983, Remark 2), if $a_{jn} = E(X_{(i)})$, where $X_{(1)}, ..., X_{(n)}$ are order statistics of a random sample from a distribution that has finite $(2 + \delta)$th absolute moment for $\delta > 0$, then (14.4) holds. Hence the logrank statistic, where the scores are defined in (7.10), is asymptotically normal.

14.2.2 The conditional test

Now we are interested in the asymptotic distribution of the conditional linear rank test; i.e., the asymptotic distribution of S_n conditional on $N_A(n) = n_{An}$. Holst's Theorem (Section 13.1) can be applied to show the asymptotic normality of the conditional test. Since (13.3) is trivially true, the main conditions on the scores for asymptotic normality are, from (13.4) and (13.5), (14.4) and

$$
\frac{\sum_{j=1}^{n}(a_{jn} - \bar{a}_{jn})^2}{n} \to \gamma \text{ as } n \to \infty, \tag{14.6}
$$

where $\gamma > 0$ is a constant. Condition (14.6) holds, for example, if

$$
a_{jn} = \frac{r_{jn}}{n+1}, \tag{14.7}
$$

where $\{r_{jn}\}$ are the simple ranks.

We then apply the theorem with $X_j = (T_j + 1)/2$. Then $\theta = 1/2$, $\sigma^2(\theta) = \theta(1 - \theta)$, and $x_n/n = n_{An}/n \to \theta$. Condition (13.6) is true. For the other non-trivial condition, (13.7), one must show that for all $\epsilon > 0$, there exists a $K_\epsilon < 1$ such that, for $\epsilon \leq |t| \leq \pi$,

$$\left| E \left(e^{itT_j/2} \right) \right| \leq K_\epsilon < 1 \tag{14.8}$$

for all j. Since

$$E \left(e^{itT_j/2} \right) = \cos(t/2),$$

then for all $\epsilon > 0$, there exists a $K_\epsilon = \cos(\epsilon/2)$ such that, for $\epsilon \leq |t| \leq \pi$,

$$\left| E \left(e^{itT_j/2} \right) \right| \leq \cos(\epsilon/2) = K_\epsilon < 1$$

for all j. Thus (14.8) is true.

We conclude from Holst's Theorem that, if (14.4) and (14.6) hold, then the conditional distribution of

$$\frac{S_n}{\sqrt{4\gamma n_{An}(n - n_{An})/n}}, \tag{14.9}$$

given $N_A(n) = n_{An}$, is asymptotically standard normal.

14.2.3   Simulation results

Simulations are based on population models, and therefore one might hesitate to discuss the simulation of tests computed under a randomization model. In fact, for a given set of scores from a data set, an appropriate simulation procedure would be to generate m replicates of n treatment assignments under the particular randomization strategy and then compute the test statistic with the *observed* scores, for a total of m simulated test statistics. But one would be interested in properties of the test for more general sets of scores, and to simulate these properties, one must generate the scores from some population-based mechanism. We have simulated the test for $m = 10,000$ replications assuming that the scores arose from ranking responses from either a standard normal or a Cauchy distribution. One could argue that it is more in keeping with the principles of randomization by taking one set of scores from a single pass simulation and m sets of treatment assignments. We generated $10,000$ sets of both in our simulations, principally so we could compare with a population-based test, the standard Student's t-test.

Table 14.1 presents simulation results from complete randomization using the unconditional test in (14.3) and the conditional test in (14.9). Ten thousand test statistics using simple ranks were generated with $n = 50$ and the tail probabilities of the standard normal distribution are reported. One can see that the sample size is certainly large enough to conclude that the test statistic is indeed standard normal,

Table 14.1   Tail probabilities of the normal distribution for the unconditional (U) and conditional (C) linear rank test (LRT) with simple rank scores and the t-test under complete randomization, $n = 50$. Results based on $10,000$ simulations with responses generated from standard normal and Cauchy distributions.

			Left Tail				Right Tail			
Test	U/C	Dist.	0.005	0.025	0.050	0.100	0.100	0.050	0.025	0.005
LRT	U	Normal	0.004	0.024	0.054	0.106	0.103	0.052	0.024	0.004
LRT	U	Cauchy	0.006	0.024	0.047	0.098	0.101	0.050	0.025	0.005
LRT	C	Normal	0.003	0.024	0.050	0.100	0.099	0.050	0.023	0.005
LRT	C	Cauchy	0.005	0.023	0.047	0.097	0.098	0.048	0.023	0.005
t-test		Normal	0.006	0.029	0.055	0.109	0.105	0.054	0.028	0.006
t-test		Cauchy	0.001	0.013	0.031	0.089	0.098	0.036	0.013	0.001

although the tests are slightly conservative in the extreme tails. The size of the test is between 0.047 and 0.049 in all cases. The test is invariant to the underlying distribution of responses.

The t-test is clearly only appropriate with standard normal responses, as one would expect. We report results with Cauchy response just to illustrate how inappropriate the test is in non-normal settings. With normal responses, the test tends to have slightly inflated size (0.059). One can conclude the the linear rank test will provide a slightly conservative normal test for samples as small as 50, and the test is invariant to any underlying distribution of the patient responses.

14.3   RANDOM ALLOCATION RULE

Asymptotic normality of the linear rank test when a random allocation rule is used can be demonstrated using Hájek's Theorem (Section 13.1). It turns out that (14.3) holds under the identical condition as for complete randomization.

Consider a statistic of the form in (13.8). In our case, we let $c_{jn} = a_{jn} - \bar{a}_n$, $u_n = 0$, $v_n = 1$ and

$$\psi(t) = \begin{array}{ll} -1, & 0 < t \le 1/2; \\ 1, & 1/2 < t < 1. \end{array}$$

We immediately see that ψ is nondecreasing on $(0, 1)$, $\bar{\psi} = 0$, and $\int_0^1 (\psi(t) - \bar{\psi})^2 dt = 1$. Then

$$\left\{ \psi\left(\frac{R_{1n}}{n+1}\right), ..., \psi\left(\frac{R_{nn}}{n+1}\right) \right\} \tag{14.10}$$

Table 14.2   Tail probabilities of the normal distribution for the linear rank test (LRT) with simple rank scores and the t-test under the random allocation rule, $n = 50$. Results based on 10,000 simulations with responses generated from standard normal and Cauchy distributions.

		Left Tail				Right Tail			
Test	Dist.	0.005	0.025	0.050	0.100	0.100	0.050	0.025	0.005
LRT	Normal	0.005	0.026	0.053	0.106	0.105	0.050	0.024	0.004
LRT	Cauchy	0.005	0.024	0.045	0.098	0.102	0.051	0.024	0.004
t-test	Normal	0.008	0.030	0.057	0.105	0.104	0.054	0.027	0.006
t-test	Cauchy	0.001	0.011	0.032	0.098	0.094	0.032	0.012	0.001

has a uniform distribution on all possible permutations of $n/2$ 1's and $n/2$ -1's. Therefore, (14.10) has the same distribution as $(T_1, ..., T_n)$ under the random allocation rule. Hence

$$S_n = \sum_{j=1}^{n}(a_{jn} - \bar{a}_n)T_j = \sum_{j=1}^{n}(a_{jn} - \bar{a}_n)\psi\left(\frac{R_{jn}}{n+1}\right).$$

Then under condition (14.4), we have from Hájek's Theorem that

$$\frac{S_n}{\sqrt{\text{Var}(S_n)}} \to N(0,1)$$

in law, as $n \to \infty$. By (13.9), using (7.13), we conclude that, if condition (14.4) holds, then

$$\frac{\sum_{j=1}^{n}(a_{jn} - \bar{a}_n)T_j}{\sqrt{\frac{n}{n-1}\sum_{j=1}^{n}(a_{jn} - \bar{a}_n)^2}}$$

is asymptotically standard normal, which is asymptotically equivalent to (14.3).

Table 14.2 gives simulation results for the random allocation rule, under the same setting as Table 14.1. One can see that the results are very similar to complete randomization, and that the linear rank test has size between 0.048 and 0.05 and that the t-test under normal response has inflated size of 0.057.

14.4   TRUNCATED BINOMIAL DESIGN

Finding the asymptotic form and distribution of the linear rank statistic for truncated binomial randomization is an open problem. We can see by simulation if the statistic is asymptotically normal ignoring the randomization procedure. In Table 14.3, we simulate the linear rank test under truncated binomial randomization, assuming the complete randomization variance in (14.3) The linear rank test does not deviate from

Table 14.3 Tail probabilities of the normal distribution for the linear rank test (LRT) with simple rank scores and the t-test under the truncated binomial design, n = 50, and variance computed assuming complete randomization. Results based on 10, 000 simulations with responses generated from standard normal and Cauchy distributions.

		Left Tail				Right Tail		
Test Dist.	0.005	0.025	0.050	0.100	0.100	0.050	0.025	0.005
LRT Normal	0.005	0.027	0.052	0.107	0.105	0.051	0.026	0.006
LRT Cauchy	0.004	0.025	0.051	0.105	0.107	0.056	0.026	0.005

normality too much; it is slightly anticonservative with the size of the test 0.051 and 0.053. Hence the variance of the test is slightly smaller than it should be, which accounts for our ignoring the extreme correlation in the tail of the truncated binomial.

14.5 EFRON'S BIASED COIN DESIGN

The biased coin design does not necessarily produce an asymptotically normal linear rank statistic, even under condition (14.4) (Smythe and Wei, 1983). For a simple counterexample, for n even, let

$$a_{jn} = 1, \quad \text{if } 1 \le j \le n/2;$$
$$= -1, \quad \text{if } n/2 + 1 \le j \le n.$$

Then $(a_{jn} - \bar{a}_n)^2 = 1$ and

$$\frac{\max_{1 \le j \le n} (a_{jn} - \bar{a}_n)^2}{\sum_{j=1}^{n}(a_{jn} - \bar{a}_n)^2} = \frac{1}{n} \to 0$$

as $n \to \infty$. Then

$$S_n = \sum_{j=1}^{n}(a_{jn} - \bar{a}_n)T_j$$

$$= \sum_{j=1}^{n/2}T_j - \sum_{j=n/2+1}^{n} T_j$$

$$= 2D_{n/2} - D_n,$$

say, where $|D_n|$ is a recurrent Markov chain on the nonnegative integers. As shown in Section 3.6, $|D_n|$ converges to a stationary distribution on the nonnegative integers as $n \to \infty$. Therefore, S_n does not have a normal limit.

Table 14.4 Tail probabilities of the normal distribution for the unconditional linear rank test (LRT) with simple rank scores (calculated with complete randomization variance) and the t-test under the biased coin design ($p = 2/3$), $n = 50$. Results based on 10,000 simulations with responses generated from standard normal and Cauchy distributions.

		\multicolumn{4}{c}{Left Tail}	\multicolumn{4}{c}{Right Tail}
Test	Dist.	0.005	0.025	0.050	0.100	0.100	0.050	0.025	0.005
LRT	Normal	0.005	0.024	0.054	0.106	0.109	0.055	0.027	0.006
LRT	Cauchy	0.005	0.026	0.055	0.105	0.102	0.052	0.025	0.004
t-test	Normal	0.007	0.029	0.056	0.104	0.108	0.057	0.030	0.007
t-test	Cauchy	0.001	0.013	0.033	0.098	0.092	0.035	0.013	0.002

For the conditional test, Hollander and Peña (1988) noted that the exact permutation test for $n = 50$ and 70 demonstrated non-normal behavior for Efron's biased coin design. They conjecture that the conditional linear rank test has a non-normal limit, and leave this as an open problem. Halpern and Brown (1986) simulated the behavior of the chi-square statistic for binary responses following randomization using Efron's biased coin design. They found that the traditional analysis ignoring the randomization procedure generally was satisfactory, except in cases where the binary responses had long sequences of runs, in which case the traditional analysis was conservative.

While the unconditional test is not always normal, at least for the linear rank test with simple rank scores, simulations in Table 14.4 suggest that it is. We used the naive variance from complete randomization in the simulations. The linear rank test appears to be normal with the correct size. Proving this rigorously is an open problem. Again the t-test has inflated size.

14.6 WEI'S URN DESIGN

Smythe and Wei (1983) derive the asymptotic distribution of the unconditional linear rank test for Wei's $UD(\alpha, \beta)$ design. The key of the proof is to equate S_n with a martingale difference array and then apply the martingale central limit theorem. The form of the test statistic is then

$$\frac{S_n}{\sqrt{\sum_{j=1}^{n} b_{jn}^2}}, \tag{14.11}$$

where $\{b_{jn}\}$ is a modified score sequence, to be defined below in (14.14).

We use this new set of constants $\{b_{jn}\}, j = 1, ..., n$ to define a new test statistic

$$
\begin{aligned}
W_{kn} &= \sum_{j=1}^{k} b_{jn} \left(T_j - E(T_j | \mathcal{F}_{j-1})\right) \\
&= \sum_{j=1}^{k} b_{jn} \left(T_j - f(j-1) \sum_{i=1}^{j-1} T_i\right),
\end{aligned} \tag{14.12}
$$

where $f(j) = -\beta/(2\alpha + \beta j)$, by (3.13). Then $W_{kn}, k = 1, ..., n$ is a martingale array (see Section 13.3).

We wish to choose the constant sequence $\{b_{jn}\}$ so that $S_n = W_{nn}$. Then applying the martingale central limit theorem to W_{nn} will show the asymptotic normality of S_n. We can write the equations

$$
\sum_{j=1}^{n} (a_{jn} - \bar{a}_n) T_j = \sum_{j=1}^{n} b_{jn} \left(T_j - f(j-1) \sum_{i=1}^{j-1} T_i\right)
$$

as

$$
\Gamma_{n \times n} B_n = A_n, \tag{14.13}
$$

where $B_n = \{b_{1n}, ..., b_{nn}\}'$, $A_n = \{(a_{1n} - \bar{a}_n), ..., (a_{nn} - \bar{a}_n)\}'$, and

$$
\Gamma = \begin{bmatrix}
1 & f(1) & f(2) & \cdots & f(n-2) & f(n-1) \\
 & 1 & f(2) & \cdots & f(n-2) & f(n-1) \\
 & & 1 & \cdots & f(n-2) & f(n-1) \\
 & & & \ddots & \vdots & \vdots & \vdots \\
0 & & & 1 & f(n-2) & f(n-1) \\
 & & & & 1 & f(n-1) \\
 & & & & & 1
\end{bmatrix}.
$$

The unique solution is obtained by inverting Γ, noting the relationship $1 + f(j) = f(j)/f(j-1)$. We obtain

$$
\begin{aligned}
b_{nn} &= a_{nn} - \bar{a}_n, \\
b_{jn} &= (a_{jn} - \bar{a}_n) + \sum_{k=j+1}^{n} \frac{f(k-1)f(k-2)}{f(j-1)} (a_{kn} - \bar{a}_n) \\
&= (a_{jn} - \bar{a}_n) \\
&\quad - \beta(2\alpha + (j-1)\beta) \sum_{k=j+1}^{n} \frac{(a_{kn} - \bar{a}_n)}{(2\alpha + (k-1)\beta)(2\alpha + (k-2)\beta)}, \\
& \qquad j = 1, ..., n-1
\end{aligned} \tag{14.14}
$$

(Problem 14.5).

We can now apply Hall and Heyde's Theorem for martingale arrays (Section 13.3) to $W_{kn}/\sqrt{\sum_{j=1}^{n} b_{jn}^2}$. Assume the following:

$$\lim_{n\to\infty} \frac{\max_{1\leq j\leq n} b_{jn}^2}{\sum_{j=1}^{n} b_{jn}^2} = 0. \tag{14.15}$$

Then conditions (13.17) and (13.19) follow immediately. For condition (13.18), we must show that

$$\frac{\sum_{j=1}^{n} b_{jn}^2 \left(T_j + \frac{\beta}{2\alpha+(j-1)\beta}\sum_{k=1}^{j-1} T_k\right)^2}{\sum_{j=1}^{n}(b_{jn})^2} \to 1 \tag{14.16}$$

in probability. This follows since $\sum_{j=1}^{n} T_j/n \to 0$ in probability by (3.22), and hence, for large j,

$$\left(T_j + \frac{\beta}{2\alpha+(j-1)\beta}\sum_{k=1}^{j-1} T_k\right)^2 \sim 1.$$

Hence (14.16) holds. We conclude from (13.20) that if (14.15) holds,

$$\frac{W_{kn}}{\sqrt{\sum_{j=1}^{n} b_{jn}^2}}$$

is asymptotically standard normal, and hence (14.11) converges in law to a standard normal distribution as $n \to \infty$.

However, we desire an easily verified condition on the original score sequence $\{a_{jn}\}$ rather than on the transformed sequence $\{b_{jn}\}$. While Smythe and Wei (1983) were unable to show the equivalence of (14.4) and (14.15), it was later shown by Wei, Smythe, and Smith (1986). The proof is not difficult, but rather tedious. The interested reader is referred to the paper for details. It is interesting to note that the variance of the test statistic in (14.11) is quite complicated compared to that of complete randomization. Smythe and Wei (1983, Remark 6) report a simulation study, using an uncentered version of the linear rank statistic, in which significance levels were considerably overestimated if the analysis of a $UD(\alpha, \beta)$ design is performed using a linear rank test with the variance computed assuming complete randomization.

Table 14.5 gives simulation results for the $UD(0, 1)$ under the same settings as Tables 14.1–14.4. However, we also compare the linear rank test using the correct variance against the test using the naive variance from complete randomization. The evidence here shows that they are almost identical, and this calls into question whether it is necessary to use the more complicated scores based on the $\{b_{jn}\}$ sequence. Table 14.6 repeats the results for $n = 25$ and finds that the size of the test is 0.49 for the correct variance and 0.56 for the complete randomization variance, somewhat validating their result. We suspect that for moderate to large samples,

Table 14.5   *Tail probabilities of the normal distribution for the unconditional linear rank test (LRT) with simple rank scores (with the variance computed assuming complete randomization and urn randomization) and the t-test under $UD(0, 1)$ randomization, $n = 50$. Results based on $10,000$ simulations with responses generated from standard normal and Cauchy distributions.*

Test	Var	Dist.	Left Tail				Right Tail			
			0.005	0.025	0.050	0.100	0.100	0.050	0.025	0.005
LRT	urn	Normal	0.005	0.023	0.048	0.101	0.103	0.046	0.024	0.004
LRT	urn	Cauchy	0.005	0.025	0.049	0.099	0.096	0.046	0.023	0.004
LRT	complete	Normal	0.005	0.025	0.051	0.104	0.107	0.049	0.026	0.005
LRT	complete	Cauchy	0.006	0.026	0.052	0.103	0.098	0.049	0.025	0.005
t-test		Normal	0.006	0.027	0.054	0.104	0.108	0.052	0.027	0.006
t-test		Cauchy	0.002	0.013	0.035	0.099	0.092	0.032	0.012	0.001

Table 14.6   *Tail probabilities of the normal distribution for the unconditional linear rank test (LRT) with simple rank scores (with the variance computed assuming complete randomization and urn randomization) under $UD(0, 1)$ randomization, $n = 25$. Results based on $10,000$ simulations with responses generated from standard normal distribution.*

Var	Left Tail				Right Tail			
	0.005	0.025	0.050	0.100	0.100	0.050	0.025	0.005
urn	0.003	0.021	0.047	0.099	0.097	0.054	0.028	0.005
complete	0.004	0.024	0.053	0.101	0.098	0.057	0.032	0.006

the two variances are virtually equivalent, with small differences evident at smaller sample sizes. Again the t-test has inflated size, and the linear rank tests seem to perform very well for $n = 50$.

The conditional test for the $UD(\alpha, \beta)$ is a special case of the conditional tests derived for the family of generalized biased coin designs, which we will discuss in the next section.

14.7   WEI, SMYTHE, AND SMITH'S GENERAL ALLOCATION RULES

14.7.1   The unconditional test for $K > 2$ treatments

The asymptotic distribution of the unconditional linear rank test for the general K-treatment allocation rules defined in (3.21) was derived in Wei, Smythe, and Smith (1986) using techniques similar to those used in Section 14.6. The principal assumption on the function p is that it be twice continuously differentiable with bounded second derivatives.

For a constant ϕ, define

$$\frac{\partial p_k(\xi)}{\partial y_i} = \delta_{ki}\phi, \tag{14.17}$$

where δ_{ki} is the Kronecker delta. Define the sequence of modified scores to be

$$b_{jn} = (a_{jn} - \bar{a}_n) + \phi \sum_{i=j+1}^{n} \frac{a_{in} - \bar{a}_n}{i - 1} \prod_{k=j}^{i-2} \left(1 + \frac{\phi}{k}\right), \tag{14.18}$$

(where by convention $\prod_{k=j}^{l} = 1$ if $l < i$). Note when $K = 2$ for the $UD(0,1)$, (14.18) reduces to (14.14), as

$$\phi = p'(1/2) = -1.$$

Let $T_{ij} = 1$ if treatment i is assigned to patient j, and $T_{ij} = 0$ otherwise (note that the centering no longer applies). Then we can define the K-treatment linear rank statistic as $S_n = (S_{1n}, ..., S_{Kn})'$, where

$$S_{in} = \frac{\sum_{j=1}^{n}(a_{jn} - \bar{a}_n)(T_{ij} - \xi_i)}{\sqrt{\sum_{j=1}^{n} b_{jn}^2}}. \tag{14.19}$$

Then we have the following result. If (14.4) holds, then

$$T_n = S_n' \Sigma^{-1} S_n \to \chi^2(K - 1), \tag{14.20}$$

where Σ has elements $\sigma_{ii} = \xi_i(1 - \xi_i)$ and $\sigma_{ik} = -\xi_i\xi_k$, $k \neq i$. The proof is similar to that of Wei and Smythe (1983), but carries with it the resultant complications of moving from $K = 2$ to general K.

14.7.2   The conditional test for two treatments

When $K = 2$, Smythe (1988) proves the analogous result for the conditional linear rank test, using a theorem of Heckman (1985). Asymptotic normality seems to require a stronger condition than (14.4). In fact, it is sufficient that

$$\frac{n \max_{1 \leq j \leq n} (a_{jn} - \bar{a}_n)^2}{\sum_{j=1}^{n} (a_{jn} - \bar{a}_n)^2} \leq C \tag{14.21}$$

for each n, where C is a constant. Smythe (personal communication) conjectures that (14.21) is not necessary, and can probably be relaxed to (14.4); this is an open problem. Smythe's proof assumes that the scores $\{a_{jn}\}$ are scaled so that

$$\sum_{j=1}^{n} b_{jn}^2 = 1,$$

where $\{b_{jn}\}$ is defined in (14.18). In practice, since $\{b_{jn}\}$ is a linear combination of $\{a_{jn} - \bar{a}_n\}$, we can compute $\kappa = \sum_{j=1}^{n} b_{jn}^2$ and then scale a_{jn} by taking

$$\frac{a_{jn} - \bar{a}_n}{\sqrt{\kappa}}.$$

One also needs to define another sequence of modified scores, denoted $\{\tilde{b}_{jn}\}$, which are computed by substituting $a_{jn} - \bar{a}_n = n^{-1/2}$ for all j into (14.18). For example, for the $UD(0,1)$, from (14.14), we have

$$\tilde{b}_{jn} = n^{-1/2} \left(1 - (j-1) \sum_{k=j+1}^{n} \frac{1}{(k-1)(k-2)} \right)$$

$$= n^{-1/2} \left(\frac{j-1}{n-1} \right) \tag{14.22}$$

(Problem 14.6). Define

$$\rho_n = \sum_{j=1}^{n} b_{jn} \tilde{b}_{jn}, \quad s^2 = \lim_{n \to \infty} \sum_{j=1}^{n} \tilde{b}_{jn}^2.$$

From (14.22), for the $UD(0,1)$, $s^2 = 1/3$ (Problem 14.6). We require the following additional condition:

$$\limsup_{n} \rho_n^2 < s^2. \tag{14.23}$$

The conditional central limit theorem will be conditional on $D_n = N_A(n) - N_B(n) = d_n = 2n_{An} - n$, where $\{d_n\}$ is a sequence of integers with the property that

$$d_n - n \text{ is even and } d_n = xn^{1/2} + o(n^{1/2}), \tag{14.24}$$

for integer x. Then

$$\frac{S_n}{\sqrt{1 - \rho_n^2/s^2}} - \frac{\rho_n x/s^2}{\sqrt{1 - \rho_n^2/s^2}}, \tag{14.25}$$

conditional on $D_n = d_n$, is asymptotically standard normal. In practice, one substitutes $d_n/\sqrt{n}$ for x.

Table 14.7   Tail probabilities of the normal distribution for the conditional linear rank test (LRT) with simple rank scores under $UD(0,1)$ randomization, $n = 50$. Results based on 10,000 simulations with responses generated from standard normal and Cauchy distributions. The last line is computed using the test statistic in (14.9), assuming complete randomization.

	Left Tail				Right Tail			
Dist.	0.005	0.025	0.050	0.100	0.100	0.050	0.025	0.005
Normal	0.005	0.022	0.047	0.100	0.101	0.046	0.024	0.004
Cauchy	0.005	0.025	0.048	0.098	0.094	0.045	0.023	0.004
Normal	0.006	0.028	0.054	0.108	0.111	0.051	0.028	0.005

Table 14.7 simulates the conditional test for the $UD(0,1)$, and can be directly compared to Table 14.5. The conditional and unconditional tests have similar properties for $n = 50$. However, if we ignore the more complicated form of the test statistic in (14.25) and instead use the form of the conditional test under complete randomization in (14.9), the test is anti-conservative, as seen in the last line of Table 14.7. Hence, unlike the unconditional test under $UD(0,1)$ randomization, we cannot use a simpler form of the test statistic ignoring the randomization when computing the conditional test statistic.

14.8   CONCLUSIONS

Table 14.8 gives the different forms of the asymptotic variance for the various randomization procedures, as well as conditions for asymptotic normality. One can see that for complete randomization and the random allocation rule, the tests are computed in the same way. It is when we come to more complicated randomization procedures, such as Wei's urn design, that the variance is more complicated, but simulation results indicate that it may not be necessary to use the more complicated form of the variance, except for the conditional test.

14.9   PROBLEMS

14.1 Show that condition (14.4) does not hold when
(i) $a_{jn} = q^j, 0 < q \neq 1$;
(ii) $a_{jn} = 1/j$. (Hájek, 1969)

14.2 Show that condition (14.4) holds for the van der Waarden scores, defined in Problem 7.4, at a rate $O(\ln n/n)$.
(Hints:
(a) Use the approximation to the normal distribution function $1 - \Phi(x) \sim \phi(x)/x$ as

Table 14.8 *Conditions required for asymptotic normality and the form of the denominator of S_n for conditional (C) and unconditional (U) linear rank tests under the various randomization procedures.*

Randomization Procedure	C/U	Conditions	Denominator^2
Complete	U	(14.4)	$\sum_{j=1}^{n}(a_{jn} - \bar{a}_n)^2$
Complete	C	(14.4), (14.6)	$4\gamma n_{An}(n - n_{An})/n$
Random allocation	same	(14.4)	$\sum_{j=1}^{n}(a_{jn} - \bar{a}_n)^2$
Truncated binomial	same	open problem	
Biased coin	U	can be non-normal	
Biased coin	C	conjectured non-normal	
$UD(\alpha, \beta)$	U	(14.4)	$\sum_{j=1}^{n} b_{jn}^2$
$UD(0, 1)$	C	(14.21), (14.23), (14.24)	$1 - \rho_n^2/s^2$

$x \to \infty$, where $\phi(x)$ is the normal density function (*e.g.*, Feller (1968, p. 175));
(b) Use the approximation $\bar{a}_n \sim \int_0^1 \Phi^{-1}(u)du$.)

14.3 Prove or give a counterexample to the following statement: "Any continuous function of the simple ranks satisfies (14.4)."

14.4 Show that, for the simple rank scores given in (14.7), the value of γ in (14.6) is 12.

14.5 Show that (14.14) is the solution to (14.13).

14.6 For the $UD(0, 1)$ conditional test, verify (14.22) and that $s^2 = 1/3$.

14.10 REFERENCES

FELLER, W. (1968). *An Introduction to Probability Theory and Its Applications.* Wiley, New York.

HALPERN, J. AND BROWN, B. W. (1986). Sequential treatment allocation procedures in clinical trials – with particular attention to the analysis of results for the biased coin design. *Statistics in Medicine* **5** 211–229.

HECKMAN, N. (1985). A local limit theorem for a biased coin design for sequential tests. *Annals of Statistics* **13** 785–788.

HOLLANDER, M. AND PEÑA, E. (1988). Nonparametric tests under restricted treatment-assignment rules. *Journal of the American Statistical Association* **83** 1144–1151.

KALISH, L. A. AND BEGG, C. B. (1987). The impact of treatment allocation on nominal significance level and bias. *Controlled Clinical Trials* **8** 121–135.

SMYTHE, R. T. (1988). Conditional inference for restricted randomization designs. *Annals of Statistics* **16** 1155–1161.

SMYTHE, R. T. AND WEI, L. J. (1983). Significance tests with restricted randomization design. *Biometrika* **70** 496–500.

WEI, L. J., SMYTHE, R. T., AND SMITH, R. L. (1986). *K*-treatment comparisons with restricted randomization rules in clinical trials. *Annals of Statistics* **14** 265–274.

15

Large Sample Inference for Response-Adaptive Randomization

15.1 INTRODUCTION

In this chapter we prove the asymptotic normality of maximum likelihood estimators from a response-adaptive randomization procedure. We also explore the large sample distribution of the linear rank test under a randomization model following response-adaptive randomization.

15.2 MAXIMUM LIKELIHOOD ESTIMATION

15.2.1 Asymptotic normality of the maximum likelihood estimator: Urn models

Rosenberger, Flournoy, and Durham's Theorem (Section 13.4.4) can be used to prove the consistency and asymptotic normality of maximum likelihood estimators from a response-adaptive randomization procedure when $E(\delta_{ji}|\mathcal{F}_{i-1})$ converges almost surely to a constant. While the theorem is appropriate for continuous outcomes, we will focus on the product binomial likelihood in (11.6). In particular, this methodology applies to the generalized Friedman's urn models introduced in Chapter 10.

Using the notation of Chapter 13, the first derivative of the loglikelihood increments is given by

$$\frac{\partial \log L_i(p_1, ..., p_K)}{\partial p_j} = \frac{(T_i - p_j)\delta_{ji}}{p_j(1 - p_j)}.$$

Since $|(T_i - p_j)\delta_{ji}| \leq 1$, we immediately see that (13.27) and (13.29) hold. Taking the second derivative, which is nonzero only for $j = k$, we see that the summands is again bounded, and so (13.30) holds. It remains only to show (13.28). Again, for $j = k$, we have

$$n^{-1} \sum_{i=1}^{n} E\left(\left(\frac{\partial L_i(p_1, \ldots, p_K)}{\partial p_j}\right)^2 \bigg| \mathcal{F}_{i-1}\right)$$

$$= n^{-1} p_j^{-2}(1 - p_j)^{-2} \sum_{i=1}^{n} E((T_i - p_j)^2 \delta_{ji} | \mathcal{F}_{i-1}).$$

Using a conditioning argument, we can show that

$$E((T_i - p_j)^2 \delta_{ji} | \mathcal{F}_{i-1}) = p_j(1 - p_j) E(\delta_{ji} | \mathcal{F}_{i-1}) \tag{15.1}$$

(Problem 15.1).

Consider the generalized Friedman's urn model. Since

$$E(\delta_{ji} | \mathcal{F}_{i-1}) \to v_j$$

almost surely, for large i, by Athreya and Karlin (1967), where v_j is defined in (10.8), we have that

$$n^{-1} \sum_{i=1}^{n} E\left(\left(\frac{\partial L_i(p_1, \ldots, p_K)}{\partial p_j}\right)^2 \bigg| \mathcal{F}_{i-1}\right) \to v_j / p_j(1 - p_j)$$

almost surely, as $n \to \infty$. By (13.32), we conclude that $\hat{p} = (\hat{p}_1, \ldots, \hat{p}_K)$ is consistent for $p = (p_1, \ldots, p_K)$ and that

$$n^{-1/2}(\hat{p} - p) \to N(0, \Sigma) \tag{15.2}$$

in law, where Σ is a diagonal matrix with elements $\sigma_{jj} = p_j(1 - p_j)/v_j$. Note that $\hat{p}_1, \ldots, \hat{p}_K$ are asymptotically independent. This is the same result we would obtain under an independent sampling scheme if $(v_1, \ldots, v_K)$ were fixed in advance. We can therefore set up the usual chi-square tests of $K - 1$ treatment comparisons to a control, or other suitable multivariate tests.

An extension to covariate-adjusted models, under the framework of generalized linear models, is straightforward but messy. See Rosenberger and Hu (2001) for details.

15.2.2 Delayed response

Bai, Hu, and Rosenberger (2002) evaluated the effects of delayed response on the asymptotic distribution of the maximum likelihood estimators. Let the entry time of the nth be denoted τ_n and assume the entry times have independent increments; that is, $\{\tau_n - \tau_{n-1}\}$ are independent and identically distributed for all n. Let the response time of the nth patient be denoted $r_n(j, l), j = 1, ..., K, l = 1, 2$, where j indexes the treatment assignment and l indexes the response. Assume that $r_n(j, l)$ has distribution g_{jl}, so that the time to response distribution can potentially depend on both the treatment assigned and the patient's response.

The question arises, under what conditions on τ_n and $r_n(j, l)$ will the limiting results in (10.8) still hold? It turns out that one needs only ensure that the probability that m additional patients will arrive prior to a patient's response is of order $o(m^{-c})$ for $c \in (0, 1]$. While this result cannot be easily verified in practice, it satisfies our intuition that the delay cannot be very large relative to the entry stream. In practice, we can verify this by examining the following conditions on τ_n and $r_n(j, l)$:

(i) $E(r_n(j, l))^{c_1} < \infty$ for each j, l, and $c_1 > c$.
(ii) $E(\tau_i - \tau_{i-1}) > 0$ and $E(\tau_i - \tau_{i-1})^2 < \infty$.

Assuming these two conditions hold, the maximum likelihood estimators will have the same asymptotic distribution as in (15.2) because the likelihood under the population model is the same as in (11.5) (Problem 15.2).

15.2.3   Likelihood ratio test for K treatments

For binary response trials of K treatments, one would like a test of the null hypothesis $H_0 : p_1 = \cdots p_K = p_0$ versus an alternative that at least one differs. For the birth and death urn, Ivanova, Rosenberger, Durham, *et al.* (2000) derive the asymptotic distribution of the likelihood ratio test statistic, given by

$$l_n = \frac{\mathcal{L}_n(\tilde{p}_n)}{\mathcal{L}_n(\hat{p}_n)},$$

where $\tilde{p}_n$ is the maximum likelihood estimator under the null hypothesis and $\hat{p}_n$ is the maximum likelihood estimator under the whole parameter space. They prove that $-2 \log l_n$ has the usual asymptotic chi-square distribution on $K - 1$ degrees of freedom under the null hypothesis. They also derive the noncentrality parameter under the alternative hypothesis, but can only prove the result for $p_0 < 1/2$. The distribution under the alternative for $p_0 > 1/2$ is an open problem.

15.2.4   Asymptotic properties of sequential maximum likelihood procedures

Melfi and Page (2000) provide a very powerful result on strong consistency of parameter estimators from a response-adaptive randomization procedure for two treatments. This result is particularly useful for the sequential maximum likelihood procedure, but can also be used for urn designs as well. Following a general adaptive randomization procedure, let $\tilde{\theta}_A$ and $\tilde{\theta}_B$ be some estimators of θ_A and θ_B, respectively, where θ_A is the parameter of interest for treatment A and θ_B is the parameter of interest for treatment B. Suppose the following two conditions are true:

(1) If the treatment responses were actually generated as a sequence of independent and identically distributed random variables, then $\tilde{\theta}_A$ and $\tilde{\theta}_B$ would be strongly consistent for θ_A and θ_B, respectively.
(2) Under the adaptive allocation procedure, $N_A(n) \to \infty$ and $N_B(n) \to \infty$ almost surely, as $n \to \infty$.

Then for any adaptive randomization procedure, $\tilde{\theta}_A$ and $\tilde{\theta}_B$ are strongly consistent for θ_A and θ_B, respectively.

While this is a result of major importance, condition (2) is not typically easy to prove. In the simplest case with binary responses, let ρ be of the form $f(p_A)/(f(p_A) + f(p_B))$, for $f : [0,1] \to [0,\infty)$. We will show that $N_A(n) \to \infty$ almost surely if

$$f(\hat{p}_B) \leq C < \infty \tag{15.3}$$

for constant C. More generally, we could also impose

$$\sup_{0 \leq x \leq 1} f(x) \leq C < \infty.$$

We leave the proof for $N_B(n)$ to the interested reader. First, we must ensure that $f(\hat{p}_A(k))$ is never zero, where $\hat{p}_A(k) = S_A(k)/N_A(k)$ and $\hat{p}_B(k) = S_B(k)/N_B(k)$ are the maximum likelihood estimators of p_A and p_B after $k = 1, ..., n$ patients. In general, this will not be the case if $f(0) = 0$, as the number of successes, $S_A(k)$ and $S_B(k)$, have positive probability of being 0, as do the number of patients allocated, $N_A(k)$ and $N_B(k)$, In practice, we would add some small constant to $S_A(k)$, $S_B(k)$, $N_A(k)$, and $N_B(k)$ to account for this. Melfi, Page, and Geraldes (2001) suggest taking $S_A(k) + 0.5$, $S_B(k) + 0.5$, $N_A(k) + 1$, and $N_B(k) + 1$.

To prove the result, note that $N_A(k)$ is nondecreasing in k. Consider the complement of the divergent set. We have

$$\{N_A(k) \to \infty\}^c = \bigcup_{m=1}^{\infty} \{N_A(j) = N_A(m) \; \forall j > m\},$$

which imples

$$\Pr\left(\{N_A(k) \to \infty\}^c\right) \leq \sum_{m=1}^{\infty} \Pr\left(\{N_A(j) = N_A(m) \; \forall j > m\}\right). \tag{15.4}$$

Let $U_1, U_2, ...$ be a sequence of uniform random numbers on $[0,1]$. Then

$$\Pr\left(\{N_A(j) = N_A(m) \; \forall j > m\}\right) \quad = \quad \Pr(U_j \geq \hat{\rho}_j(p_A, p_B) \; \forall j > m)$$

$$= \quad \Pr\left(U_j \geq \frac{f(\hat{p}_A(j-1))}{f(\hat{p}_A(j-1)) + f(\hat{p}_B(j-1))} \; \forall j > m\right),$$

$$= \quad \Pr\left(U_j \geq \frac{\hat{p}_A(m)}{\hat{p}_A(m) + \hat{p}_B(j-1)} \; \forall j > m\right),$$

since $f(\hat{p}_A)(m) = f(\hat{p}_A)(j)$ on the set in question. Now by (15.3), we have

$$\Pr\left(U_j \geq \frac{\hat{p}_A(m)}{\hat{p}_A(m) + \hat{p}_B(j-1)} \;\forall j > m\right)$$

$$\leq \Pr\left(U_j \geq \frac{\hat{p}_A(m)}{\hat{p}_A(m) + C} > 0 \;\forall j > m\right)$$
$$= 0.$$

By (15.4), this implies that

$$\Pr\left(\{N_A(k) \to \infty\}^c\right) = 0,$$

and hence $N_A(k) \to \infty$ almost surely, as $k \to \infty$.

In addition, under conditions (1) and (2) above, Melfi, Page, and Geraldes (2001) prove that

$$\frac{N_A(n)}{n} \to \rho(\theta_A, \theta_B)$$

almost surely, as $n \to \infty$. This result can then be used to prove asymptotic normality of the maximum likelihood estimators using the techniques in Section 13.4.

15.3 LARGE SAMPLE LINEAR RANK TESTS

As far as we know, there has been no work on asymptotic properties of linear rank tests for covariate-adaptive randomization and only one paper deriving the large sample distribution of the linear rank test for response-adaptive randomization. This would certainly be an fertile area for future research.

Rosenberger (1993) gives the form of the permutation test for the randomized play-the-winner rule. The basic idea is similar to developing a large sample permutation test for Wei's urn design, given in Section 14.6. As in Chapter 14, let $T_i = 1$ if treatment A is assigned and $T_i = -1$ if treatment B is assigned, and $a_{in} = 1$ if the treatment was a success and $a_{in} = 0$ if failure. We wish to prove the asymptotic normality of the test statistic

$$S_n = \sum_{i=1}^{n}(a_{in} - \bar{a}_n)T_i$$

under $RPW(\alpha, 1)$ randomization. To do this, we equate S_n to a martingale by selecting a sequence of scores $b_{in}, i = 1, ..., n$ so that the martingale array formed by

$$W_{nk} = \sum_{i=1}^{k} b_{in}a_{in}(T_i - E(T_i|\mathcal{F}_{i-1}))$$

satisfies $W_{nn} = S_n$. In this case, from (10.13), we have

$$W_{nn} = \sum_{i=1}^{n} b_{in} a_{in} \left(T_i - \frac{\sum_{j=1}^{i-1} a_{jn} T_j}{2\alpha - j + 1} \right).$$

Equating like terms of b_{in}, we see that

$$b_{nn} = 1,$$
$$b_{in} = \prod_{j=i+1}^{n} \left(1 + \frac{a_{jn}}{2\alpha + i - 1} \right), \qquad (15.5)$$

$i = 1, ..., n - 1$ (Problem 15.5). We now employ Hall and Heyde's Theorem for martingale arrays (Section 13.3) to show that W_{nn}, suitably normalized, and hence S_n, is asymptotically standard normal.

Under the following two conditions,

$$\lim_{n \to \infty} \frac{\max_{1 \le i \le n} b_{in}^2}{\sum_{i=1}^{n} b_{in}^2} = 0 \qquad (15.6)$$

and

$$E\left((T_i - E(T_i|\mathcal{F}_{i-1})^2|\mathcal{F}_{i-1}) \right) \to 1 \qquad (15.7)$$

in probability, as $i \to \infty$, we have that

$$\frac{S_n}{\sqrt{\sum_{i=1}^{n} b_{in}^2}} \to N(0, 1) \qquad (15.8)$$

in law, as $n \to \infty$. Note that conditions (13.17), (13.19), along with (13.21) are all satisfied under condition (15.6). It is easier to show (13.22) than (13.18), and it results immediately from (15.6) and (15.7). (See Problem 15.6.) Now let us examine condition (15.7) more carefully. We can rewrite the condition as

$$(E(T_i|\mathcal{F}_{i-1}))^2 \to 0$$

in probability, as $i \to \infty$, or that

$$\frac{\sum_{j=1}^{n} a_{jn} T_j}{n} \to 0 \qquad (15.9)$$

in probability, as $n \to \infty$.

It is not possible to ensure that (15.6) and (15.9) will hold for all possible observed sequences of a_{jn}. In fact, if all $a_{jn} = 1$, neither condition holds; $\sum_{j=1}^{n} a_{jn} T_j/n$ in fact converges to a beta random variable almost surely (*e.g.*, Athreya and Ney (1972, p. 220)). At the other extreme, if all $a_{jn} = 0$, asymptotic normality of S_n is well known (*e.g.*, Freedman (1965)). To determine how a "typical" response sequence might behave, Rosenberger (1993) assumes that the a_{jn} arose as Bernoulli random variables with parameter p. If $p < 3/4$, he shows that both conditions (15.6) and (15.7) hold. Consequently, if the sequence of responses can be considered to be a realization from a Bernoulli sequence, provided $p < 3/4$, we can conclude that (15.8) holds.

Rosenberger (1993) also explored the large sample distribution of a linear rank statistic for continuous outcomes, following response-adaptive randomization using a treatment effect mapping of the linear rank statistic outlined in Section 10.6. The test is more complicated than that for the randomized play-the-winner rule, and the reader is referred to Rosenberger (1993) for details.

15.4 PROBLEMS

15.1 Verify (15.1).

15.2 Verify that the likelihood under the delayed response model in Section 15.2.2 is identical to that when there is immediate response. Let $\boldsymbol{\tau}^{(j)} = (\tau_1, ..., \tau_j)$ be the entry times and $\boldsymbol{r}^{(j)} = (r_1, ..., r_j)$ be the response times of patients $1, ..., j$. Show, by appropriate conditioning that

$$\mathcal{L}_n = \mathcal{L}(\boldsymbol{r}^{(n)}, \boldsymbol{y}^{(n)}, \boldsymbol{t}^{(n)}, \boldsymbol{z}^{(n)}, \boldsymbol{\tau}^{(n)}; \boldsymbol{\theta})$$

reduces to (11.5).

15.3 Use Melfi and Page's approach in Section 15.2.4 to show that the maximum likelihood estimators $\hat{p}_A$ and $\hat{p}_B$ from the randomized play-the-winner rule are strongly consistent.

15.4 Consider a sequential maximum likelihood randomization procedure for binary response designed to target an allocation

$$\rho(p_A, p_B) = \frac{\frac{1}{\sqrt{p_A q_A}}}{\frac{1}{\sqrt{p_A q_A}} + \frac{1}{\sqrt{p_B q_B}}},$$

the optimal allocation for minimizing expected failures for the odds ratio measure. Let $\theta = p_A q_B / p_B q_A$ be the odds ratio with estimator $\hat{\theta}_n = \hat{p}_A \hat{q}_B / \hat{p}_B \hat{q}_A$. Find the asymptotic distribution of $\sqrt{n}(\hat{\theta}_n - \theta)$.

15.5 Derive the scores in (15.5).

15.6 Show that (15.6) and (15.7) imply (13.21) and (13.22).

15.5 REFERENCES

ATHREYA, K. B. AND KARLIN, S. (1968). Embedding of urn schemes into continuous time Markov branching processes and related limit theorems. *Annals of Mathematical Statistics* **39** 1801–1817.

ATHREYA, K. B. AND NEY, P. (1972). *Branching Processes*. Springer, Berlin.

BAI, Z. D., HU, F., AND ROSENBERGER, W. F. (2002). Asymptotic properties of adaptive designs for clinical trials with delayed response. *Annals of Statistics* **30** 122-139.

FREEDMAN, D. (1965). Bernard Friedman's urn. *Annals of Mathematical Statistics* **36** 956–970.

IVANOVA, A. V., ROSENBERGER, W. F., DURHAM, S. D., AND FLOURNOY, N. (2000). A birth and death urn for randomized clinical trials: asymptotic methods. *Sankhya B* **62** 104–118.

MELFI, V. F. AND PAGE, C. (2000). Estimation after adaptive allocation. *Journal of Statistical Planning and Inference* **87** 353–363.

MELFI, V. F., PAGE, C., AND GERALDES, M. (2001). An adaptive randomized design with application to estimation. *Canadian Journal of Statistics* **29** 107–116.

ROSENBERGER, W. F. (1993). Asymptotic inference with response-adaptive treatment allocation designs. *Annals of Statistics* **21** 2098–2107.

ROSENBERGER, W. F., FLOURNOY, N., AND DURHAM, S. D. (1997). Asymptotic normality of maximum likelihood estimators from multiparameter response-driven designs. *Journal of Statistical Planning and Inference* **60** 69–76.

ROSENBERGER, W. F. AND HU, M. (2001). On the use of generalized linear models following a sequential design. *Statistics and Probability Letters* **56** 155–161.

Author Index

Subject Index

WILEY SERIES IN PROBABILITY AND STATISTICS
ESTABLISHED BY WALTER A. SHEWHART AND SAMUEL S. WILKS

Editors: *David J. Balding, Peter Bloomfield, Noel A. C. Cressie,*
Nicholas I. Fisher, Iain M. Johnstone, J. B. Kadane, Louise M. Ryan,
David W. Scott, Adrian F. M. Smith, Jozef L. Teugels
Editors Emeriti: *Vic Barnett, J. Stuart Hunter, David G. Kendall*

The *Wiley Series in Probability and Statistics* is well established and authoritative. It covers many topics of current research interest in both pure and applied statistics and probability theory. Written by leading statisticians and institutions, the titles span both state-of-the-art developments in the field and classical methods.

Reflecting the wide range of current research in statistics, the series encompasses applied, methodological and theoretical statistics, ranging from applications and new techniques made possible by advances in computerized practice to rigorous treatment of theoretical approaches.

This series provides essential and invaluable reading for all statisticians, whether in academia, industry, government, or research.

*Now available in a lower priced paperback edition in the Wiley Classics Library.

*Now available in a lower priced paperback edition in the Wiley Classics Library.

*Now available in a lower priced paperback edition in the Wiley Classics Library.

*HOAGLIN, MOSTELLER, and TUKEY · Understanding Robust and Exploratory
    Data Analysis
HOCHBERG and TAMHANE · Multiple Comparison Procedures
HOCKING · Methods and Applications of Linear Models: Regression and the Analysis
    of Variables
HOEL · Introduction to Mathematical Statistics, *Fifth Edition*
HOGG and KLUGMAN · Loss Distributions
HOLLANDER and WOLFE · Nonparametric Statistical Methods, *Second Edition*
HOSMER and LEMESHOW · Applied Logistic Regression, *Second Edition*
HOSMER and LEMESHOW · Applied Survival Analysis: Regression Modeling of
    Time to Event Data
HØYLAND and RAUSAND · System Reliability Theory: Models and Statistical Methods
HUBER · Robust Statistics
HUBERTY · Applied Discriminant Analysis
HUNT and KENNEDY · Financial Derivatives in Theory and Practice
HUSKOVA, BERAN, and DUPAC · Collected Works of Jaroslav Hajek—
    with Commentary
IMAN and CONOVER · A Modern Approach to Statistics
JACKSON · A User's Guide to Principle Components
JOHN · Statistical Methods in Engineering and Quality Assurance
JOHNSON · Multivariate Statistical Simulation
JOHNSON and BALAKRISHNAN · Advances in the Theory and Practice of Statistics: A
    Volume in Honor of Samuel Kotz
JUDGE, GRIFFITHS, HILL, LÜTKEPOHL, and LEE · The Theory and Practice of
    Econometrics, *Second Edition*
JOHNSON and KOTZ · Distributions in Statistics
JOHNSON and KOTZ (editors) · Leading Personalities in Statistical Sciences: From the
    Seventeenth Century to the Present
JOHNSON, KOTZ, and BALAKRISHNAN · Continuous Univariate Distributions,
    Volume 1, *Second Edition*
JOHNSON, KOTZ, and BALAKRISHNAN · Continuous Univariate Distributions,
    Volume 2, *Second Edition*
JOHNSON, KOTZ, and BALAKRISHNAN · Discrete Multivariate Distributions
JOHNSON, KOTZ, and KEMP · Univariate Discrete Distributions, *Second Edition*
JUREČKOVÁ and SEN · Robust Statistical Procedures: Aymptotics and Interrelations
JUREK and MASON · Operator-Limit Distributions in Probability Theory
KADANE · Bayesian Methods and Ethics in a Clinical Trial Design
KADANE AND SCHUM · A Probabilistic Analysis of the Sacco and Vanzetti Evidence
KALBFLEISCH and PRENTICE · The Statistical Analysis of Failure Time Data
KASS and VOS · Geometrical Foundations of Asymptotic Inference
KAUFMAN and ROUSSEEUW · Finding Groups in Data: An Introduction to Cluster
    Analysis
KENDALL, BARDEN, CARNE, and LE · Shape and Shape Theory
KHURI · Advanced Calculus with Applications in Statistics
KHURI, MATHEW, and SINHA · Statistical Tests for Mixed Linear Models
KLUGMAN, PANJER, and WILLMOT · Loss Models: From Data to Decisions
KLUGMAN, PANJER, and WILLMOT · Solutions Manual to Accompany Loss Models:
    From Data to Decisions
KOTZ, BALAKRISHNAN, and JOHNSON · Continuous Multivariate Distributions,
    Volume 1, *Second Edition*
KOTZ and JOHNSON (editors) · Encyclopedia of Statistical Sciences: Volumes 1 to 9
    with Index
KOTZ and JOHNSON (editors) · Encyclopedia of Statistical Sciences: Supplement
    Volume

*Now available in a lower priced paperback edition in the Wiley Classics Library.

*Now available in a lower priced paperback edition in the Wiley Classics Library.

MYERS, MONTGOMERY, and VINING · Generalized Linear Models. With Applications in Engineering and the Sciences

NELSON · Accelerated Testing, Statistical Models, Test Plans, and Data Analyses

NELSON · Applied Life Data Analysis

NEWMAN · Biostatistical Methods in Epidemiology

OCHI · Applied Probability and Stochastic Processes in Engineering and Physical Sciences

OKABE, BOOTS, SUGIHARA, and CHIU · Spatial Tesselations: Concepts and Applications of Voronoi Diagrams, *Second Edition*

OLIVER and SMITH · Influence Diagrams, Belief Nets and Decision Analysis

PANKRATZ · Forecasting with Dynamic Regression Models

PANKRATZ · Forecasting with Univariate Box-Jenkins Models: Concepts and Cases

*PARZEN · Modern Probability Theory and Its Applications

PEÑA, TIAO, and TSAY · A Course in Time Series Analysis

PIANTADOSI · Clinical Trials: A Methodologic Perspective

PORT · Theoretical Probability for Applications

POURAHMADI · Foundations of Time Series Analysis and Prediction Theory

PRESS · Bayesian Statistics: Principles, Models, and Applications

PRESS and TANUR · The Subjectivity of Scientists and the Bayesian Approach

PUKELSHEIM · Optimal Experimental Design

PURI, VILAPLANA, and WERTZ · New Perspectives in Theoretical and Applied Statistics

PUTERMAN · Markov Decision Processes: Discrete Stochastic Dynamic Programming

*RAO · Linear Statistical Inference and Its Applications, *Second Edition*

RENCHER · Linear Models in Statistics

RENCHER · Methods of Multivariate Analysis, *Second Edition*

RENCHER · Multivariate Statistical Inference with Applications

RIPLEY · Spatial Statistics

RIPLEY · Stochastic Simulation

ROBINSON · Practical Strategies for Experimenting

ROHATGI and SALEH · An Introduction to Probability and Statistics, *Second Edition*

ROLSKI, SCHMIDLI, SCHMIDT, and TEUGELS · Stochastic Processes for Insurance and Finance

ROSENBERGER and LACHIN · Randomization in Clinical Trials: Theory and Practice

ROSS · Introduction to Probability and Statistics for Engineers and Scientists

ROUSSEEUW and LEROY · Robust Regression and Outlier Detection

RUBIN · Multiple Imputation for Nonresponse in Surveys

RUBINSTEIN · Simulation and the Monte Carlo Method

RUBINSTEIN and MELAMED · Modern Simulation and Modeling

RYAN · Modern Regression Methods

RYAN · Statistical Methods for Quality Improvement, *Second Edition*

SALTELLI, CHAN, and SCOTT (editors) · Sensitivity Analysis

*SCHEFFE · The Analysis of Variance

SCHIMEK · Smoothing and Regression: Approaches, Computation, and Application

SCHOTT · Matrix Analysis for Statistics

SCHUSS · Theory and Applications of Stochastic Differential Equations

SCOTT · Multivariate Density Estimation: Theory, Practice, and Visualization

*SEARLE · Linear Models

SEARLE · Linear Models for Unbalanced Data

SEARLE · Matrix Algebra Useful for Statistics

SEARLE, CASELLA, and McCULLOCH · Variance Components

SEARLE and WILLETT · Matrix Algebra for Applied Economics

SEBER · Linear Regression Analysis

SEBER · Multivariate Observations

SEBER and WILD · Nonlinear Regression

*Now available in a lower priced paperback edition in the Wiley Classics Library.

SENNOTT · Stochastic Dynamic Programming and the Control of Queueing Systems

*SERFLING · Approximation Theorems of Mathematical Statistics

SHAFER and VOVK · Probability and Finance: It's Only a Game!

SMALL and McLEISH · Hilbert Space Methods in Probability and Statistical Inference

STAPLETON · Linear Statistical Models

STAUDTE and SHEATHER · Robust Estimation and Testing

STOYAN, KENDALL, and MECKE · Stochastic Geometry and Its Applications, *Second Edition*

STOYAN and STOYAN · Fractals, Random Shapes and Point Fields: Methods of Geometrical Statistics

STYAN · The Collected Papers of T. W. Anderson: 1943–1985

SUTTON, ABRAMS, JONES, SHELDON, and SONG · Methods for Meta-Analysis in Medical Research

TANAKA · Time Series Analysis: Nonstationary and Noninvertible Distribution Theory

THOMPSON · Empirical Model Building

THOMPSON · Sampling, *Second Edition*

THOMPSON · Simulation: A Modeler's Approach

THOMPSON and SEBER · Adaptive Sampling

TIAO, BISGAARD, HILL, PEÑA, and STIGLER (editors) · Box on Quality and Discovery: with Design, Control, and Robustness

TIERNEY · LISP-STAT: An Object-Oriented Environment for Statistical Computing and Dynamic Graphics

TSAY · Analysis of Financial Time Series

UPTON and FINGLETON · Spatial Data Analysis by Example, Volume II: Categorical and Directional Data

VAN BELLE · Statistical Rules of Thumb

VIDAKOVIC · Statistical Modeling by Wavelets

WEISBERG · Applied Linear Regression, *Second Edition*

WELSH · Aspects of Statistical Inference

WESTFALL and YOUNG · Resampling-Based Multiple Testing: Examples and Methods for p-Value Adjustment

WHITTAKER · Graphical Models in Applied Multivariate Statistics

WINKER · Optimization Heuristics in Economics: Applications of Threshold Accepting

WONNACOTT and WONNACOTT · Econometrics, *Second Edition*

WOODING · Planning Pharmaceutical Clinical Trials: Basic Statistical Principles

WOOLSON and CLARKE · Statistical Methods for the Analysis of Biomedical Data, *Second Edition*

WU and HAMADA · Experiments: Planning, Analysis, and Parameter Design Optimization

YANG · The Construction Theory of Denumerable Markov Processes

*ZELLNER · An Introduction to Bayesian Inference in Econometrics

ZHOU, OBUCHOWSKI, and McCLISH · Statistical Methods in Diagnostic Medicine

*Now available in a lower priced paperback edition in the Wiley Classics Library.

Learning Resources